T0327686

Advances in Agricultural Systems Modeling

Transdisciplinary Research, Synthesis, and Applications

Volume 9

Lajpat R. Ahuja, Series Editor

EDITORS

Dennis J. Timlin, Saseendran S. Anapalli

CONTRIBUTORS

Samuel G. K. Adiku, Department of Soil Science, School of Agriculture, College of Basic and Applied Sciences University of Ghana, Legon, Accra, Ghana; **Shakir Ali**, Principal Scientist, Indian Institute of Soil and Water Conservation (IISWC) Research Centre, Kota (Rajasthan) India; **Saseendran S. Anapalli**, USDA_ARS Sustainable Water Management Research Unit, Stoneville, MS; **Aleme Araya**, Department of Agronomy, Kansas State University, Manhattan, KS; **Celine Birnholz**, International Center for Tropical Agriculture (CIAT), Nairobi, Kenya; **Ignacio A. Ciampitti**, Department of Agronomy, Kansas State University, Manhattan, KS; **Bright S. Freduah**, Soil and Irregation Research Centre, University of Ghana, Kapong, Ghana; **Xiaoyu Gao**, Water Conservancy and Civil Engineering College, Inner Mongolia Agricultural University, Hohhot, 010018, China; **Prasanna H. Gowda**, USDA-ARS, Southeast Area, Stoneville, MS; **Zailin Huo**, Center for Agricultural Water Research in China, China Agricultural University, Beijing, 100083, China; **Adlul Islam**, Natural Resource Division, Indian Council of Agricultural Research, New Delhi, India; **Abdullah A. Jaradat [Deceased]**, Research Agronomist, USDA-ARS, North-Central Soil Conservation Research Laboratory, Morris, MN; **Seungtaek Jeong**, Applied Plant Science, Chonnam National University, Gwangju, Republic of Korea; **Alpha Y. Kamara**, International Institute of Tropical Agriculture, R4D Unit, Ibaden, Nigeria; **Han-Yong Kim**, Applied Plant Science, Chonnam National University, Gwangju, Republic of Korea; **Jonghan Ko**, Applied Plant Science, Chonnam National University, Gwangju, Republic of Korea; **Joseph Kugbe**, Department of Soil Science, University of Development Studies, Nyankpala, Ghana; **Byunwoo Lee**, Plant Science, Seoul National University, Seoul, Republic of Korea; **Dilys S. MacCarthy**, Soil and Irregation Research Centre, School of Agriculture, College of Basic and Applied Sciences, University of Ghana, Kapong, Ghana; **Ravic Nijbroek**, International Center for Tropical Agriculture (CIAT), Nairobi, Kenya; **Prabhat R. Ojasvi**, ICAR-Indian Institute of Soil and Water Conservation Research Centre, Rajasthan, India; **Birthe Paul**, International Center for Tropical Agriculture (CIAT), Nairobi, Kenya; **P.V. Vara Prasad**, Department of Agronomy, Kansas State University; Feed the Future Innovation Lab for Collaborative Research on Sustainable Intensification, Kansas State University, Manhattan, KS; **Zhongyi Qu**, Water Conservancy and Civil Engineering College, Inner Mongolia Agricultural University, Hohhot, China; **Charles W. Rice**, Department of Agronomy, Kansas State University, Manhattan, KS; **Rolf Sommer**, International Center for Tropical Agriculture (CIAT), Nairobi, Kenya; **Pengcheng Tang**, Institute of Water Resources for Pastoral Area, China Institute of Water Resources and Hydropower Research, Hohhot, 010020, China; **Dennis J, Timlin**, USDA-ARS Adaptive Cropping Systems Laboratory, Henry A. Wallace Agricultural Research Center, Beltsville, MD

EDITORIAL CORRESPONDENCE

American Society of Agronomy
Crop Science Society of American
Soil Science Society of America
5585 Guilford Road, Madison, WI 53711-58011, USA

Advances in Agricultural Systems Modeling

Transdisciplinary Research, Synthesis, and Applications

Volume 9

Lajpat R. Ahuja, Series Editor

Enhancing Agricultural Research and Precision Management for Subsistence Farming by Integrating System Models with Experiments

Dennis J. Timlin, Saseendran S. Anapalli, Editors

Copublication by American Society of Agronomy, Inc. / Crop Science Society of America, Inc. / Soil Science Society of America, Inc. and John Wiley & Sons, Inc.

Editorial Correspondence:
American Society of Agronomy, Inc.
Crop Science Society of America, Inc.
Soil Science Society of America, Inc.
5585 Guilford Road, Madison, WI 53711-58011, USA

agronomy.org • crops.org • soils.org

Registered Offices:
John Wiley & Sons, Inc., 111 River Street, Hoboken, NJ 07030, USA

For details of our global editorial offices, customer services, and more information about Wiley products, visit us at www.wiley.com.

Wiley also publishes its books in a variety of electronic formats and by print-on-demand. Some content that appears in standard print versions of this book may not be available in other formats.

Library of Congress Cataloging-in-Publication Data applied for
Paperback 9780891183907
doi:10.1002/9780891183891

Cover Design: Wiley
Cover Image: © Thannaree Deepul/Shutterstock

Set in 9/11pt Palatino by Straive, Pondicherry, India

10 9 8 7 6 5 4 3 2 1

This book is dedicated to the memory of Dr. Abdullah Jaradat who passed away in 2020 while the book was being prepared. Dr. Jaradat was an ARS Research Agronomist at the USDA-Agricultural Research Service-North Central Soil Conservation Research Laboratory (Soil's Lab) in Morris, MN. Dr. Jaradat was recognized nationally and internationally as an authority on biodiversity, genetic resources, salinity, and cropping systems under dryland, irrigated, continental and desert agroecosystems. He was a passionate scientist, curious and eager to learn new tools for managing, manipulating, and analyzing big data. One of the recent tools Dr. Jaradat learned to use were crop models. His interest was to understand constraints to production in the dryland agricultural systems of the Fertile Crescent, one of the birth places of agriculture around the world. The result is an excellent chapter on an assessment of all the factors that affect the productivity of dryland systems in this area and impacts from climate change.

USDA-ARS Soil Management Research Unit

1

Dennis J. Timlin
and Saseendran S. Anapalli

Introduction: System Models Integrated with Experiments Can Be Useful Tools to Develop Improved Management Practices for Subsistence Farming to Address Increased Intensification and Climate Change

Semi-arid to sub-humid regions of the world are major producers of food and fiber. Soil, water, and climate resources are becoming major limiting factors for agriculture in these regions, especially in developing countries where subsistence farming is dominant. Driving factors include increased urban and industrial use of land, more frequent droughts, climate warming, and natural limits to precious natural resources. At the same time, there is a need to produce even more food for the growing population, which requires more intensive use of these resources. To obtain the most production from available resources while maintaining environmental quality, we need whole-system based quantitative knowledge and tools to help select appropriate crops and optimally manage water and associated soil inputs at different locations on a site-specific basis under variable and changing climate. Site-specific experimental results are available for limited locations, limited periods of time, and limited management options. Well-tested, process models of cropping systems can extend field research results to long-term weather conditions, other climates, and soils (Beah et al., 2021; Falconnier et al., 2020). This will allow us to explore new management options, and thus provide the whole-system based knowledge and management guides for various locations

Enhancing Agricultural Research and Precision Management for Subsistence Farming by Integrating System Models with Experiments, First Edition. Edited by Dennis J. Timlin and Saseendran S. Anapalli.

DOI: 10.1002/9780891183891.ch01

over variable weather and climate. The contributions to this book present various applications of crop system models to help develop management decision tools to optimize the use of limited water and soil nutrients for subsistence farming (low-input agriculture) and explore adaptations and mitigations for climate change under a variety of conditions. Subsistence farming generally involves smaller farms and the use of suboptimal amounts of nitrogen, water, and other inputs. The lower inputs require the models to more rigorously account for soil nitrogen–water–temperature interactions than in the cases where adequate amounts of nitrogen and water are applied.

Chapter 2 (Birnholz et al., 2022) covers the use of models to assess sustainable land management practices in the Central Highlands of Kenya. This multi-faceted study used a whole farm model to investigate impact of farm configurations (such as physiographic location, socio-economics, and crop and livestock choices) on farm performance. The information from this model was used to evaluate the outcomes of various erosion control practices on farm health and erosion. They found that profitability trade-offs existed for erosion control practices. In some cases, labor investment for maintenance of erosion control practices is not always compensated by improved income from the new practice. Also, livestock can sometimes take better advantage of improved land management practices rather than row crops.

There is some uncertainty about the best management practices for wheat (*Triticum aestivum* L.) on the sandy loam soils in Ethiopia, especially in light of potential future climate change. Wheat is a major cereal crop in Ethiopia and yields need to be higher to support a growing population. However, poor soil conditions and biotic stresses limit potential yields. Chapter 3 (Araya et al., 2022) analyzes the power of crop simulation models to better understand what management factors can be used to increase wheat yields in these soils of Ethiopia. Such models can be used to augment more expensive and time consuming field-scale experiments. The simulation results indicate that early planting with a plant density of 300 plants m^{-2} is optimal for this region and that increased nitrogen fertilization to a maximum of 180 kg ha^{-1} increases yields. The expected increase in average temperature along with increased CO_2 due to climate change in this region, according to simulations, will increase wheat yields. The analysis did not take short episodes of extreme temperatures into account.

Agro-ecosystem services provide agricultural, societal, and ecological benefits to all humans. These services include recharge of groundwater, climate modification, and maintenance of soil health, among other services. Climate change threatens the sustainability of these services. Chapter 5 (Ali et al., 2022) describes various simulation and numerical methods in detail used to quantify infiltration, aquifer recharge, and soil erosion in India and other nearby countries. These methods are then applied to understand how expected changes in climate will impact ecosystem services. Climate change models predict the wet season to become wetter and the dry season to become drier. Simulations show how water harvesting and storing excess summer water in ponds can increase groundwater recharge and provide additional water during dry periods. Because of increased future rainfall, soil erosion is predicted to increase by 58% and soil organic carbon loss is predicted to increase by 57% over the base period.

Chapter 7 (Ko et al., 2022) reports on a geospatial modeling approach to understand regional and geographical variations in rice (*Oryza sativa* L.) and barley (*Hordeum vulgare* L.) yields in Korea in the present and future environments. The simulations show that rice yields are more negatively affected by rising temperatures than is barley yield. Increased CO_2 has little effect on alleviating the effects of high temperature on rice, whereas it has a larger effect on barley. Because the climate and geography of Korea varies greatly, there is great opportunity to identify areas where climate change could have beneficial effects such as bringing warmer temperatures to higher elevations. This will allow planners to mitigate some effects of climate change by adjusting cropping methods to geography. Here the authors developed statistical methods to divide Korea into regions that similarly respond to climate change. Such a classification will help agricultural planners better target areas for what will respond favorably to climate change.

China is a country with a large population and sizable areas where water is limited during the growing season. Arid areas are especially vulnerable to climate change, and it is important to understand the hydrology and water relations of arid lands. In these arid areas, groundwater can be a significant water source for agriculture. Chapter 6 (Gao et al., 2022) discusses modeling approaches to determine the best use of groundwater that can be supplied by capillary fluxes in the soil. Shallow groundwater can provide large amounts of water via capillary rise but can limit growth due to water logging. If the groundwater is very deep, greater than 3 or 4 m upward capillary flow cannot sustain crop growth. Irrigation with water pumped from groundwater can enhance the contribution of groundwater through capillary flow and increase irrigation water use efficiency. This is because when irrigation is used in areas with a groundwater depth less than 3 m, excess irrigation that percolates to the groundwater table is available for uptake by plants. These kinds of analyses are useful in areas with plentiful groundwater but undeveloped irrigation systems.

Maize (*Zea mays* L.) is an important food crop in Africa. Adequate fertilizer can provide large yields, but there is often insufficient nitrogen fertilizer available or affordable to maintain optimum maize yields. Also, nitrogen uptake and use efficiency depends greatly on water availability, which, in turn, is affected by planting date. Chapter 4 (MacCarthy et al., 2022) explains how the Decision Support System for Agrotechnology (DSSAT) maize model is used to determine yield response to nitrogen fertilizer, variety, and sowing date at multiple locations in Ghana. The results demonstrate the spatial distribution of yield response to these variables in locations throughout Ghana and show that, at the eastern part of the Guinea Savanna, early and late planting best utilizes available rainfall and gives the best yields. Late planting is preferred in the Sudan Savanna area. They also found that, for early and very early varieties, 60 kg N ha^{-1} gave the best yields. Medium and long season varieties respond to higher nitrogen rates (90 kg ha^{-1}), and the medium season variety gives better overall yields for simulations throughout the entire country.

The Fertile Crescent (FC) in the Middle East is considered one of the birthplaces of agriculture and is also in a region where the climate has historically been highly variable and subject to drought and high temperatures. Climate change is expected to result in an increasing frequency of drought and high temperatures in this region. Because of the pressure of increasing population and intensification of agricultural activities, there is a concern of increased soil degradation and decreasing production. Chapter 8 (Jaradat & Timlin, 2022) shows how simulation models and statistical modeling are used to identify the effects of climate change on yield gaps in this area. Under the three climate change scenarios used in the study, Jaradat and Timlin (2022) found annualized crop rotation and rainfed yields were expected to decrease by 34 and 65%, respectively, compared with current conditions. In spite of using crop rotations, soil carbon is also expected to decrease due to limited rainfall and increasing temperatures. This underscores the need to develop new cropping rotations that conserve soil carbon.

Jaradat and Timlin (2022) also look at yield gaps, which is the difference between best potential yield for the area and actual yields measured on farms. They find that three-way relationships between potential yield (*Y*p), and yield gaps (*Y*g) under RCP4.5 and RCP8.5 for the major crops separated the seven countries of the FC into three groups: Jordan and Iraq are expected to sustain the largest yield gaps; Turkey, Lebanon and Iran, the lowest; and Syria and Israel intermediate yield gaps. They conclude that closing these yield gaps demands immediate local adaptive research and will inevitably involve the adoption of management practices and inputs that have been developed and used elsewhere in the world during the 1900s.

The simulation studies in this book cover a wide variety of applications. They range from assessment of fertilizer and planting date on wheat and maize yields to studies of climate change impacts on surface and subsurface water resources and erosion. Models are applied to provide insight into geospatial distributions of crop response to different environmental and management variables. We also saw examples of socio-economic analysis tied with model results. Many of the studies investigate the effects of climate change on crop yields using estimated climate data generated by climate change models (Representative Concentration Pathways RCP2.6, RCP4.5, and RCP8.5; and CMIP5 [Coupled Model Intercomparison Project, https://www.wcrp-climate.org/wgcm-cmip/wgcm-cmip5]). The effects on subsistence and low-input agriculture are expected to be severe due to lack of research on alternative management practices and limited resources for farmers. However, most studies show that increased diversification through crop rotations and use of alternative crops can decrease the variability caused by climate change and result in decreased impact of heat and water stress.

References

Ali, S., Islam, A., & Ojasvi, P. R. (2022). Modeling water dynamics for assessing and managing ecosystem services in India. In D. Timlin (Ed.), *Enhancing agricultural research and precision management for subsistence farming by integrating system models with experiments* (pp. 69–103). ASA, CSSA, and SSSA. DOI

Araya, A., Prasad, P. V. V., Ciampitti, I. A., Rice, C. W., & Gowda, P. H. (2022). Using crop simulation models as tools to quantify effects of crop management practices and climate change scenarios on wheat yields in northern Ethiopia. In D. Timlin (Ed.), *Enhancing agricultural research and precision management for subsistence farming by integrating system models with experiments* (pp. 29–47). ASA, CSSA, and SSSA. DOI

Beah, A., Kamara, A. Y., Jibrin, J. M., Akinseye, F. M., Tofa, A. I., & Adam, A. M. (2021). Simulating the response of drought–tolerant maize varieties to nitrogen application in contrasting environments in the Nigeria Savannas using the APSIM model. *Agronomy, 11*, 76. https://doi.org/10.3390/agronomy11010076

Birnholz, C., Paul, B., Sommer, R., & Nijbroek, R. (2022). Modeling soil erosion impacts and trade-offs of sustainable land management practices in the Upper Tana Region of the Central Highlands in Kenya. In D. Timlin (Ed.), *Enhancing agricultural research and precision management for subsistence farming by integrating system models with experiments* (pp. 6–28). ASA, CSSA, and SSSA. doi 10.1002/9780891183891

Falconnier, G. N., Corbeels, M., Boote, K. J., Affholder, F., Adam, M., MacCarthy, D. S., Ruane, A. C., Nendel, C., Whitbread, A. M., Justes, É., Ahuja, L. R., Akinseye, F. M., Alou, I. N., Amouzou, K. A., Anapalli, S. S., Baron, C., Basso, B., Baudron, F., Bertuzzi, P., . . . Webber, H. (2020). Modelling climate change impacts on maize yields under low nitrogen input conditions in sub-Saharan Africa. *Global Change Biology*, *26*, 5942–5964. https://doi.org/10.1111/gcb.15261

Gao, X., Huo, Z., Qu, Z., & Tang, P. (2022). Modeling agricultural hydrology and water productivity to enhance water management in the Arid Irrigation District of China. In D. Timlin (Ed.), *Enhancing agricultural research and precision management for subsistence farming by integrating system models with experiments* (pp. 104–133). ASA, CSSA, and SSSA. doi 10.1002/9780891183891

Jaradat, A. A., & Timlin, D. (2022). Constraints to productivity of subsistence dryland agroecosystems in the Fertile Crescent: Simulation and statistical modeling. In D. Timlin (Ed.), *Enhancing agricultural research and precision management for subsistence farming by integrating system models with experiments* (pp. 155–189). ASA, CSSA, and SSSA. doi 10.1002/9780891183891

Ko, J., Jeong, S., Kim, H.-Y., & Lee, B. (2022). Use of data and models in simulating regional and geospatial variations in climate change impacts on rice and barley in the Republic of Korea. In D. Timlin (Ed.), *Enhancing agricultural research and precision management for subsistence farming by integrating system models with experiments* (pp. 134–154). ASA, CSSA, and SSSA. doi 10.1002/9780891183891

MacCarthy, D. S., Adiku, S. G. K., Kamara, A. Y., Freduah, B. S., & Kugbe, J. (2022). The role of crop simulation modeling in managing fertilizer use in maize production systems in northern Ghana. In D. Timlin (Ed.), *Enhancing agricultural research and precision management for subsistence farming by integrating system models with experiments* (pp. 48–68). ASA, CSSA, and SSSA. doi 10.1002/9780891183891

2

Celine Birnholz, Birthe Paul,
Rolf Sommer,
and Ravic Nijbroek

Modeling Soil Erosion Impacts and Trade-Offs of Sustainable Land Management Practices in the Upper Tana Region of the Central Highlands in Kenya

Abstract

This study was designed to accompany the Upper Tana Nairobi Water Fund, a large public–private partnership fund that promotes sustainable land management (SLM) practices among smallholder farmers to protect supply and quality of water to one of Kenya's most economically important regions. Co-benefits or trade-offs of the promoted practices might include implications for smallholder farmers' livelihoods and climate change, which influence adoption of technologies and sustainability of interventions. We modeled the soil, greenhouse gas (GHG), and profitability impacts of three SLM practices—terrace maintenance, and Napier grass (*Pennisetum purpureum* Schumacher) and *Brachiaria* grass strips—which were selected through participatory land management evaluations with farmers. Household survey data, soil sampling and tree measurements were combined with published experimental data to calibrate the whole farm bio-economic model FarmDESIGN for six farming systems across three Nairobi Water Fund priority sub-watersheds. Results illustrate the heterogeneity of the farms in terms of types of soils, crops, livestock, and their management practices, which affect the potential impact of SLM practices. Profitability trade-offs for SLM practices exist, especially in cases where higher labor investment (e.g., for terrace maintenance) is not

Enhancing Agricultural Research and Precision Management for Subsistence Farming by Integrating System Models with Experiments, First Edition. Edited by Dennis J. Timlin and Saseendran S. Anapalli.

DOI: 10.1002/9780891183891.ch02

compensated for by other income sources (e.g., higher milk production from improved feed obtained from grass strips). Grass strips in integrated crop–livestock systems can be a key entry point for sustainable development programs, reducing erosion and providing associated co-benefits such as reduced GHG emission intensity. This study demonstrated the added value of linking bio-economic modeling and participatory socio-economic assessments for contextualized ex-ante impact assessments to inform development interventions.

Introduction

The Upper Tana River Basin covers approximately 17,000 km in Central Kenya and is home to 5.3 million people (TNC, 2015). The water it provides is of critical importance to the Kenyan economy: it fuels one of the country's most important agricultural areas, provides half of the country's hydropower output, and supplies 95% of Nairobi's water. It is also home to areas of unique biodiversity and important national parks. One of the major challenges in the Upper Tana is that upstream human activities are contributing to increased sedimentation in the basin's rivers, reducing the capacity of reservoirs and increasing the costs for water treatment. The Upper Tana Nairobi Water Fund was created in 2015 to help protect and restore the quality and supply of water to one of Kenya's most productive and economically important regions. Today, 60% of Nairobi's residents are water insecure (TNC, 2015). The challenges to water security will likely grow as climate change brings increasingly unpredictable rainfall. The Nairobi Water Fund's goals are to increase wet season flow by 30% and dry season flow by 15% (TNC, 2015). In the establishment period, the fund conducted an analysis to spatially target their investment portfolios. Soil and water conservation interventions were selected by the fund's steering committee based on a review of literature on conservation interventions (TNC, 2015). The analysis resulted in the selection of six practices to achieve the fund's objectives: management of riparian vegetation along riverbanks, adoption of agroforestry, introduction of terraces on hill slopes in steep farmland, reforestation for degraded lands on forest edges, planting grass strips in farmlands, and mitigation of erosion from dirt roads. The Nairobi Water Fund activities are currently focused in three priority sub-watersheds, and rivers from these sub-watersheds are critical to Nairobi's water supply and Kenya's power supply.

Water funds are founded on the principle that it is cheaper to prevent water problems at the source than it is to address them farther downstream. Spearheaded by The Nature Conservancy (TNC), the Nairobi Water Fund involves government, nongovernment, and research organizations, as well as businesses and utility companies. The fund has established a revolving fund, where a public–private partnership of donors and major water consumers "at the tap" contribute to the endowment, which generates financial resources to support land-conservation measures upstream. In March 2015, the Nairobi Water Fund was launched after 2 yr of groundwork, and it was Africa's first water fund. A business case to assess the economic viability of a water fund for the entire Upper Tana River Basin was conducted in 2015 and demonstrates a clear economic justification for its establishment. Although the business case made the economic argument at watershed level, it did not look at potential trade-offs at the farm level and assumed that all proposed activities can be implemented in the targeted areas.

A challenge for the Nairobi Water Fund is providing evidence that activities are resulting in the expected outcomes. Additionally, although the benefits of Nairobi Water Fund are aimed at beneficiaries downstream, ensuring that land-users upstream benefit from land-conservation measures will play an important role in the long-term viability and sustainability of the fund. The agricultural landscape is made of a network of smallholder farms that have delineated crop, pasture, and tree plots throughout the Upper Tana River Basin. To evaluate whether interventions schemes, such as by the Nairobi Water Fund, developed for an entire watershed at highly aggregated level, provide sufficient benefits and incentives to smallholder farmers, we need an assessment of how land-conservation measures, and more generally, how sustainable land management (SLM) practices, impact productivity, and livelihoods at farm level.

Ex-ante impact assessments using bio-economic farm models, such as FarmDESIGN (Groot et al., 2012), allow farm level assessments using multiple variables, and are adapted to different agro-ecological settings. FarmDESIGN has been previously used in other parts of East Africa, for example for the ex-ante impact assessment of soil protection technologies in western Kenya (Paul et al., 2015), solution spaces for nutrition-sensitive agriculture in western Kenya (Timler et al., 2020), and exploring agro-environmental trade-offs across various livestock systems in northern Tanzania (Paul et al., 2020).

This study aimed to simulate the soil erosion impacts and trade-offs of preferred SLM technologies in selected sub-watersheds of the Upper Tana region of the Central Highlands of Kenya. The specific objectives were to (a) understand current agricultural systems including soils and crop and livestock holdings and practices; (b) estimate baseline soil erosion taking into account current SLM practices that were inventoried on site; (c) quantify agro-environmental performance of farming systems in terms of nutrient balances, greenhouse gas (GHG) emissions, and profits; and (d) estimate impacts and trade-offs of selected preferred SLM practices on the same dimensions.

Materials and Methods

Study Site and Farming System Selection

The study was conducted in the Thika–Chania, Maragua, and Sagana watersheds, which are the priority sub-watersheds of the Nairobi Water Fund (Figure 2.1).

These watersheds are prone to erosion due to relatively steep slopes which increase at higher elevations, especially from 2,600 m above sea level upward (Figure 2.2).

Generally, land holdings of smallholders in this region are in one location and extend in strips from the crest of hills down into the valley bottoms to the rivers or creeks, and thus nearly all farms have access to water. In the tea zone (above 1500 m), tea (*Camellia sinensis* L.) is grown on hillsides and covers about 75% of the farm whereas vegetables and trees are grown on the crest of the hills near the homesteads and along the river bottoms (Braslow & Cordingley, 2016). In the rainy season, food crops (maize [*Zea mays* L.], bean [*Phaseolus vulgaris* L.]) are planted on the upper part of their farms in rotation, and in the dry season potato (*Solanum tuberosum* L.) and sweet potato [*Ipomoea batatas* (L.) Lam.] and various vegetables are irrigated on the lower part of farms along the rivers.

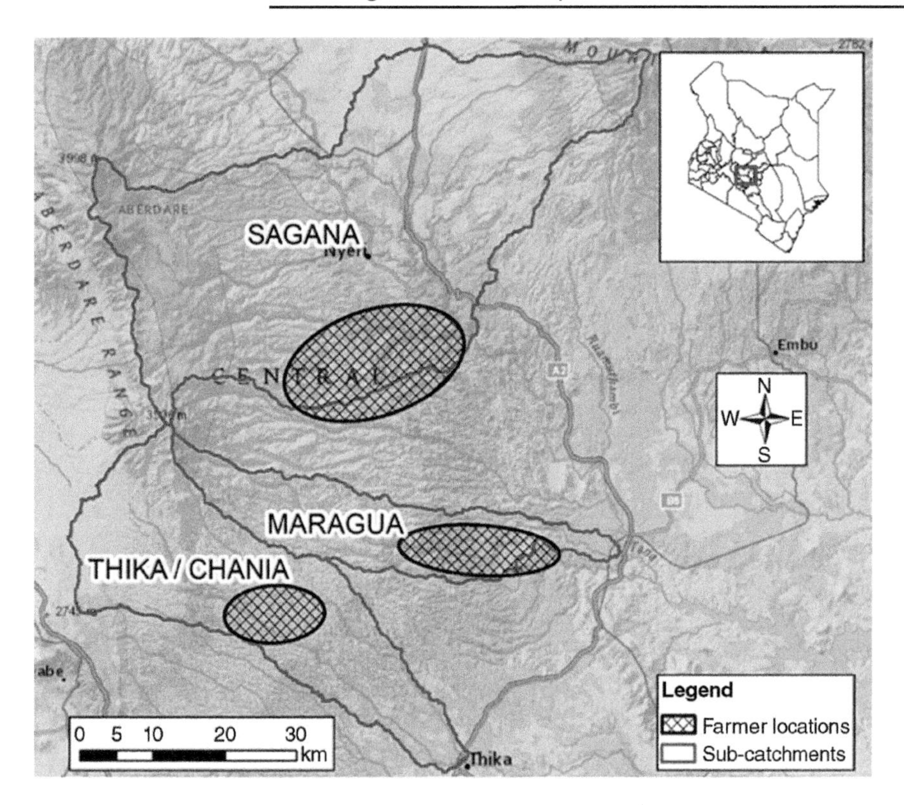

Figure 2.1 Map of the three sub-watersheds in which the modeled farms are located.

Household and Biophysical Data Collection

Six representative farms across the three sub-watersheds were selected from the study participants of Emerton and Gicheha (2016, unpublished) in discussion with extension officers: TK1 and TK2 from Thika–Chania, MR1 and MR2 from Maragua, and SG1 and SG2 from Sagana. These six farms were interviewed two times: first data collection was conducted using the ImpactLite survey tool (Rufino et al., 2012) from 16 to 19 May 2017, and model inputs and results were validated with the farmers from 4 to 8 Sept. 2017. The ImpactLite survey provided information on socio-economic attributes and livelihood strategies including characteristics of the farm household head, family structure, land (land size, plots, and tenure), crop production (crops, yields, utilization of crop products and residues, income from sales, labor activities, and inputs), livestock production (herd composition, production, income from sales, labor activities, feeding, management, and inputs), and off farm activities (income).

In short, the six farming systems can all be characterized as small mixed crop–livestock farms with diverse on-farm production. Farms had between 6 and 12 different plots spread across varying slopes. All farmers kept improved dairy cattle and grew maize and bean mainly for household consumption, with some farms growing cash crops for sale. Farm TK1 was characterized as a "traditional

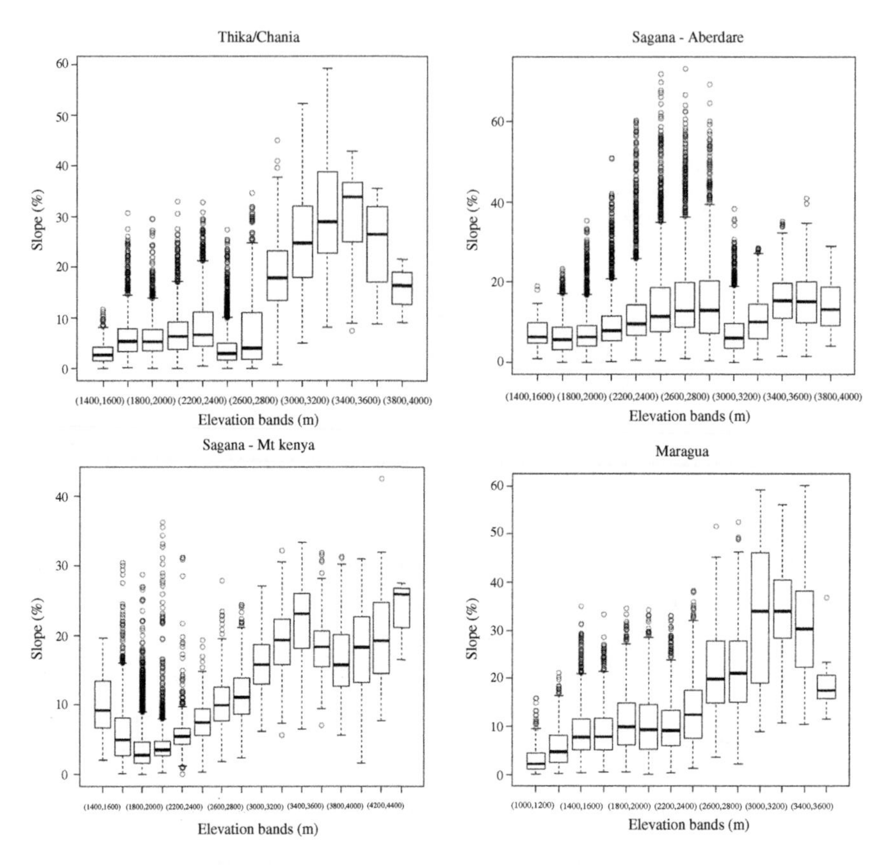

Figure 2.2 Average slope (%) per elevation band (m) for the study sub-watersheds, calculated using the NASA Shuttle Radar Topography Mission (STRM) 90 m global digital elevation data. Sagana was split between Aberdares and Mt. Kenya sides.

farm with dairy production." It was the smallest farm just under 0.1 ha with crops grown for household consumption and potato, banana (*Musa* sp.), and pineapple [*Ananas comosus* (L.) Merr.] produced for sale. The farmer did not use inorganic fertilizer and relied solely on the manure from its dairy cows for fertilization. The farmer purchased some feeds (concentrates and hay) for its two dairy cows because the on-farm Napier grass (*Pennisetum purpureum* Schumach.) and naturally occurring off-farm grasses were not sufficient. Milk and some chicken were sold to the market. Farm TK2 was called a "commercially oriented cabbage, tea, and livestock farm." At 1.39 ha, it was the largest farm, of which more than half (0.85 ha) was occupied by two tea plots. In addition to tea, several cash crops were grown including maize, cabbage (*Brassica oleracea* L.), zucchini (*Cucurbita pepo* L.), potato, and arrow root [*Colocasia esculenta* (L.) Schott]. Only the tea crop received mineral fertilizers, whereas the rest of the crops were fertilized with on-farm produced manure. This farmer also has the largest livestock herd with three cows, five sheep, and a goat. Compared with other farmers, it had a relatively large-scale poultry production compared with 600 layers and 300 broilers. Farm MR1 can be characterized as a "traditional farm with dry season cash crops," and occupied a relatively small area of 0.36 ha. Maize, bean, and banana were mainly grown for two seasons. Dry season irrigated cucumber (*Cucumis sativus* L.) and chickpea (*Cicer arietinum* L.) were produced for sale. Unlike TK1 and TK2, both mineral fertilizers and on-farm manure were applied to the crops. The farmer grew Napier grass for the small herd of one dairy cow and a calf, and half of the milk production was sold. Farm MR2 was a "commercially oriented tomato farm" and was the second largest farm (1.13 ha). In addition to the main cash crop tomato, potato and cucumber were also produced for sale, and mineral fertilizers and manure were utilized for crop fertilization. Some Napier grass was planted on contours, which was used for feeding the single dairy cow and calf. The farm had a small chicken production for sale. Farm SG1 was a "traditional farm with some horticultural cash crops and tea." This was a small farm of only 0.31 ha and several crops were produced for sale, including maize, Swiss chard (*Beta vulgaris* L.), cabbage, and collard greens (*Brassica oleracea* L.). The farm had a heifer and two sheep that fed on home-grown Napier grass and grazed on small plots of grasslands. This farmer did not use any mineral fertilizers except for the tea crop. Farm SG2 was a "traditional farm with coffee" (*Coffea arabica* L.). It was the second smallest farm (0.21 ha), and coffee occupied half of the fields. Most production was for household consumption, but some crops were produced for sale, including cabbage, banana, arrowroot, and kale. Napier was grown for one dairy cow and a heifer, and half of the milk was sold.

During the farm visits, every plot recorded during the survey was visited to measure the slopes of the plots, which were recorded in degrees using a global positioning system (GPS) device. To estimate the C stock in the trees on the farm, the number of different species of trees and the diameter at breast height (DBH) of the trees were recorded. In the case where the researcher found trees that could not be measured, DBH was estimated. Aboveground biomass (AGB) was calculated using an allometric equation by Kuyah et al. (2012). The C content in the trees was estimated by multiplying the AGB by 0.48 (kg C kg DW^{-1}), which is the average C content of wood. A total of 59 composite soil samples were collected from every plot and analyzed for texture, pH, C, N, P, and K content at CropNuts (Crop Nutrition Laboratory Services Ltd.) in Nairobi. Bulk density was measured separately at a depth of 20 cm.

Baseline Erosion Calculations

The impact of different vegetative barriers (Napier grass with leguminous shrubs) planted in contour hedgerows on soil erosion in the Central Highlands of Kenya has been measured and reported by Mutegi et al. (2008). The control in this study consisted of maize plots with no hedgerows, and Table 2.1 summarizes measured soil loss at four slope categories. These erosion values, along with the degree of the slopes for every plot measured on the six farms, were used to estimate the amount of soil loss per year in this study. No soil loss was assumed on plots with slopes between 0 and 5%. Baseline soil losses were reduced in the presence of crop cover and existing land management practices that are known to reduce erosion, using

Table 2.1 Mean soil loss from control plots (maize with no erosion measures) for different slope categories

Slope category	5–10%	10–20%	20–30%	30+
		–soil loss, Mg ha^{-1}–		
First season	16.80	79.50	77.40	67.50
Second season	16.50	79.60	79.30	78.90
Total per year	33.30	159.10	156.70	146.40

Note. Adapted from Mutegi et al. (2008).

the following assumptions: erosion is reduced by 20% with banana and pineapple cover, 60% with tree cover, 90% with tea cover, 80% with Napier grass strips, 90% with fanya juu (terrace bunds and ditches along the contour, usually also planted with Napier grass), and 97% with terraces.

Bio-Economic Modeling with FarmDESIGN

We used FarmDESIGN for modeling the six farming systems. FarmDESIGN is a bio-economic whole farm model that calculates the impacts of various farm configurations on a large set of performance indicators, including nutrient balances, labor balances, and annual income (Groot et al., 2012). As input data, we used data collected during the detailed household surveys (see Materials and Methods). The inputs required for the model can be grouped into: (a) biophysical environment (e.g., soils, climate); (b) socio-economics (e.g., costs, labor price); (c) crops and crop products yield, composition, and use; (d) livestock and livestock products yield, composition, and use; (e) manure types and degradation, and mineral fertilizer use; and (f) household members and labor availability. Farm performance is evaluated in terms of livestock feed balance, organic matter (OM) balance, farm nutrient balance and cycles, and income. The model requires soil erosion levels (mm per year) as an input. This is different from, for example, a hydrologic model such as SWAT, which simulates erosion based on land cover, slope, and precipitation. Soil sample results were averaged per farm for use in FarmDESIGN. A summary of environmental input data for the model is reported in Table 2.2.

Inventory and Prioritization of Sustainable Land Management Practices

In a separate study by Emerton and Gicheha (2016, unpublished), the ELMO (Evaluating Land Management Options) tool was used to better understand the needs, preferences, opportunities, and constraints that shape people's choices between different land management decisions. The ELMO tool relies on participatory techniques to investigate these perceptions and preferences. Seven SLM practices were discussed with a group of 22 farmers from 22 villages in Maragua Constituency (formerly District) in the Upper Tana River Basin. These practices were grass strips, irrigation, legumes, manure, mineral fertilizer, terraces, and tree planting (Table 2.3).

During the second round of interviews (see Materials and Methods), farmers revealed which SLM practices were already present on farms, and which SLM practices were preferred by the farmer, regardless of whether it was currently

Table 2.2 Environmental input parameters per farm for the FarmDESIGN model

Parameters	TK1	TK3	MR1	MR2	SG1	SG2
Soil type	Sandy clay loam			Clay		
Active OM, %	0.98	0.92	0.30	0.44	0.55	0.42
OM degradation, % yr^{-1}	2.00	2.00	2.00	2.00	2.00	2.00
Soil depth, m	0.20	0.20	0.20	0.20	0.20	0.20
Bulk density, kg m^{-3}	677.00	700.00	869.00	919.00	793.00	840.00
Texture factor	1.00	.80	.80	.80	.80	.80
Soil pH	5.26	5.29	6.25	6.25	5.11	4.76
Mean temperature, °C	17.50	17.50	17.50	17.50	17.50	17.50
Period with pF<3.5, d yr^{-1}	275.00	275.00	275.00	275.00	275.00	275.00
Soil eroded, mm yr^{-1}[a]	2.91	2.29	1.63	1.20	1.75	0.03

Note. DM, dry matter; OM, organic matter.
[a] Conversion of total soil loss (Mg ha–1) to mm is based on the assumption that 1 mm of soil equals 13 Mg DM as set up in the model parameters.

Table 2.3 Summary of farmer-reported land management practice opportunities and constraints

	Opportunities and potential	Constraints and limitations
Grass strips	Reduces soil erosion and enhances soil moisture. More likely to give lasting impacts. Provides a source of fodder for livestock, generating cash income, savings in expenditures on purchased feed, or both.	Particularly high labor demands, which often exceed what is available to the farmer.
Irrigation	Higher and more certain impacts on food, income, and crop yields. More likely to give lasting impacts. Effective in helping to fill food and cash gaps at critical times of the year—can schedule and plan the timing of crops and harvests.	Relatively high demand for family labor and technical know-how. Particularly high cash demands, which often exceed what is available to the farmer. Benefits and gains can take a long time to accrue.
Legumes	Effective in helping to fill food and cash gaps at critical times of the year.	Particularly high labor and cash demands, which often exceed what is available to the farmer.
Manure	Higher and more certain impacts on food, income, crop yields, and soil fertility. More likely to give lasting impacts.	Particularly high labor demands, which often exceed what is available to the farmer. More likely to bring termites and other pests.
Mineral fertilizer	Higher and more certain impacts on crop yields.	High demand for technical know-how and bought inputs. Particularly high cash demands, which often exceed what is available to the farmer.
Terraces	Higher and more certain impacts on improving crop yields, reducing soil erosion, enhancing soil moisture and soil fertility. More likely to give lasting impacts.	High level of technical know-how, high demand for bought inputs. Particularly high labor and cash demands, which often exceed what is available to the farmer.
Tree planting	Higher and more certain impacts in reducing soil erosion and enhancing soil moisture. More likely to give lasting impacts. Provides a source of wood, fruit, and other products, generating cash income, savings in expenditures on purchased items, or both.	Relatively high demand for bought inputs and technical know-how. Benefits and gains can take a long time to accrue.

Note. Source: Emerton and Gicheha (2016, unpublished).

Table 2.4 Three most preferred SLM practices by farms modeled in this study

Farm	SLM1	SLM2	SLM3
TK1	Terraces on specific plots	Grass strips	Manure—buy or produce more to apply to the entire farm
TK2	Grass strips (already existing, but would want to make them more productive by applying manure)	Irrigation of two plots (nothing needed as already in place)	Tree planting at edge of farm
MR1	Fanya juu—maintain the existing ones	Manure—buy or produce more to apply to the entire farm	Grass strips
MR2	Irrigation—water pan	Manure—buy or produce more to apply to the entire farm	Terraces—already in place, but would need maintenance
SG1	Manure—buy or produce more to apply more	Grass strips	Terraces everywhere
SG2	Fanya juu—maintain the existing ones	Grass strips	Manure—buy or produce more

implemented or potentially implemented in the future (Table 2.4). The farmer was asked to describe how this SLM would be implemented, e.g., where on the farm, what crops would be planted, how much it would cost, and how much labor would be necessary.

Sustainable Land Management Scenario Description

Based on the findings from the ELMO study, the analysis of the Nairobi Water Fund, and farmer feedback from the six farms (Table 2.4), three scenarios were selected. These were modeled across the farms and specifically tailored to the individual farms. The six farmers already had existing infrastructures and practices in place that can impact soil erosion (especially terraces and fanya juu), although these were in different conditions. Thus some farmers' preferences were to maintain them as they must be kept in working condition to continue benefiting from the impact that these structures have on soil erosion. The first scenario modeled was therefore maintenance and improvement of existing structure, called "maintenance." Additional costs of labor for the maintenance of terraces and/or fanya juu were modeled. Labor costs were calculated according to the data collected on the individual farms. Farm TK2 was the only farm where the farmer did not mention this as a priority or a requirement. Furthermore, because the farmer currently employs a permanent worker all year long, it was mentioned that the permanent employee could provide labor for any additional activities the farm would require. The maintenance scenario was therefore not applied to TK2.

Grass strips were a preferred practice mentioned by five out of the six farmers, and Scenarios 2 and 3 revolve around them. Scenario 2 was named "NapierStrips" and employed a range of assumptions. Grass strips were planted with Napier grass yielding 12 Mg dry matter (DM) ha^{-1} yr^{-1}, and grass was harvested and fed to the livestock. Grass strips were placed on all crop plots that did not yet have grass strips except Napier plots, grass plots, tea, and coffee plots. It was also assumed that the installation of the grass strips would reduce the crop plots on which they are installed by 10%, as this amount of land was allocated to the actual grass strips.

Table 2.5 Grass strip scenarios description (NapierStrips and BrachiariaStrips)

Farm	TK1	TK2	MR1	MR2	SG1	SG2
Grass strips area, ha	0.01	0.03	0.04	0.11	0.01	0.01
% of total field area	9.30	2.00	9.80	9.50	3.10	4.20
Napier production, Mg DM/farm	0.11	0.34	0.52	1.30	0.11	0.11
Brachiaria production, Mg DM/farm	0.07	0.20	0.31	0.77	0.07	0.06
Soil loss reduction from baseline, Mg ha^{-1} yr^{-1}	0.00	12.87	0.00	0.00	13.13	0.00

Note. DM, dry matter.

Napier reduced the calculated erosion by 80% and was applied to all plots with the new grass strips. There was no change in soil erosion on plots that had other soil erosion measures in place i.e., terraces. The amount of crop land reduced and allocated to new Napier grass strips varied across the six farms based on the assumptions described above (Table 2.5).

The third scenario, named "BrachiariaStrips," assumes that grass strips were planted with the *Brachiaria* hybrid cv. Mulato II instead of Napier grass. Brachiaria was assumed to yield 7.2 Mg DM ha^{-1}. The same assumptions as in the Napier grass strips were made. In addition, existing Napier grass strips in the baselines were replanted with Brachiaria. It has been demonstrated that Brachiaria can positively impact milk production due to higher quality and palatability than other available livestock feeds (Muinga et al., 2016). Thus, it was assumed that feeding the dairy cows the grass from the strips would result in a 5% milk production yield increase since Brachiaria from strips only constituted a small portion of the livestock diet.

Results and Discussion

Soil Quality, Erosion, and N Balances

The soil results per plot for the six farms are presented in Table 2.6. All farms had clay textured soil except for TK1 (sand clay loam). The average pH was highest (6.25) in the Maragua sub-watershed and lowest (4.76) in the Sagana sub-watershed. Total C and total N were the highest for TK1 and TK2 located in the Thika–Chania sub-watershed, respectively, 4.91 g C kg^{-1} and 4.60 g N kg^{-1}. Average total C ranged from 1.48 to 2.77 g C kg^{-1} on the four other farms MR1, MR2, SG1, and SG2. The estimated C stock in trees on the farms ranged from 410 kg C (SG1) to 11,079 kg C (SG2) with a farm average of 3,202 kg C (Table 2.6).

To estimate the soil loss per farm, we first calculated the area of each field under different slope categories (Table 2.7). Overall, all farms had plots with varying levels of steepness. All farms had fields on slopes greater than 30%, ranging from 1 to 81% of total farm area. Most farm area on steep slopes were in Thika–Chania, with TK1 having 81% of the farm on steep slopes and TK2 having 76 and 15% of plots on slopes of 20–30%. Farm MR2 had 70% of its fields on 20–30% slopes and 25% of fields on 1–20% slopes. Farm SG1 had nearly 50% of plots on 20–30% slopes. Farms MR1 and SG2 were the flattest farms, with 50 and 88%, respectively, on flat areas.

We calculated the average soil loss per year based on a combination of steepness of the slopes, crop grown, and SLM practices currently in place. Estimated soil erosion ranged from 0.45 Mg ha^{-1} (SG2) to 37.88 Mg ha^{-1} (TK1) (Table 2.8).

Table 2.6 Soil characteristics per farm and per plot and carbon stock from on-farm trees

Watershed	Farm	Plot	Area	Land use	Bray P	K	pH–H_2O	Clay + silt	Total soil N	Total soil C	Total tree C stock	Total tree C stock
			ha		mg kg^{-1}	cmol$_c$ kg^{-1}		%	g N kg^{-1}	g C kg^{-1}	kg C	kg C ha^{-1}
Thika Chania	TK1	1	0.008	Bean	22.10	4.04	6.17	43.40	4.60	56.40	1,942	20,170
		2	0.004	Napier grass	6.78	0.53	5.00	–	4.00	50.30	–	–
		3	0.012	Bean	1.22	0.40	4.62	–	3.90	49.70	–	–
		4	0.01	Maize	11.00	–	–	–	3.90	51.40	–	–
		5	0.039	Maize	2.82	–	–	–	4.00	53.90	–	–
		6	0.101	Pineapple	1.54	–	–	–	2.30	35.20	–	–
		7	0.012	Napier grass	0.51	–	–	–	2.30	–	–	–
		8	0.007	Maize	19.90	–	–	–	3.00	45.30	–	–
		9	0.021	Fallow	4.30	–	–	–	3.50	58.50	–	–
		10	0.014	Kale	4.13	–	–	–	3.80	50.00	–	–
		11	0.039	Onion	2.34	–	–	–	2.50	39.20	–	–
Thika Chania	TK2	1	0.307	Napier grass	1.67	0.14	4.81	63.50	3.60	36.90	1,242	892
		2	0.035	Maize	83.90	3.09	6.49		4.90	46.10	–	–
		3	0.325	Tea	< 0.20	0.39	4.56		3.60	46.10	–	–
		4	0.331	Napier grass	1.65	–	–	–	4.40	44.50	–	–
		5	0.028	Maize	7.74	–	–	–	4.40	44.80	–	–
		6	0.047	Cabbage	20.50	–	–	–	5.20	51.30	–	–
		7	0.136	Fallow	29.60	–	–	–	5.50	–	–	–
		8	0.052	Maize	38.70	–	–	–	5.60	48.50	–	–
		9	0.172	Potato	21.80	–	–	–	5.50	50.00	–	–
		10	1.297	Tea	1.56	–	–	–	4.30	49.10	–	–
		11	0.098	Maize	19.00	–	–	–	4.70	52.20	–	–
		12	0.051	Arrow root	3.29	–	–	–	3.50	37.20	–	–

Table 2.6 (Continued)

Watershed	Farm	Plot	Area	Land use	Bray P	K	pH–H_2O	Clay + silt	Total soil N	Total soil C	Total tree C stock	Total tree C stock
			ha		mg kg^{-1}	$cmol_c$ kg^{-1}		%	g N kg^{-1}	g C kg^{-1}	kg C	kg C ha^{-1}
Maragua	MR1	1	0.242	Maize	0.28	0.42	6.47	91.50	2.70	13.70	2,953	8,204
		2	0.013	Napier grass	0.32	0.15	6.85	–	2.80	13.60	–	–
		3	0.003	Napier grass	1.90	0.14	5.42	–	3.00	12.60	–	–
		4	0.022	Maize	30.70	–	–	–	3.00	19.60	–	–
		5	0.003	Kale	2.06	–	–	–	0.80	8.20	–	–
		6	0.009	Napier grass	1.69	–	–	–	2.90	16.00	–	–
		7	0.15	Banana	1.10	–	–	–	2.80	14.60	–	–
		8	0.004	Maize	2.11	–	–	–	2.90	19.20	–	–
		9	0.285	Maize	2.19	–	–	–	2.60	18.60	–	–
		10	0.16	Maize	8.13	–	–	–	2.80	17.70	–	–
		11	NA	Maize	< 0.20	–	–	–	0.81	8.80	–	–
Maragua	MR2	1	0.141	Maize	3.24	0.75	5.59	91.50	2.80	16.80	1,589	1,400
		2	0.059	Potato	0.61	1.38	6.98	–	2.80	25.10	–	–
		3	0.043	Maize	1.01	1.22	6.17	–	2.50	21.00	–	–
		4	0.008	Cow pea	12.60	–	–	–	2.80	26.20	–	–
		5	0.015	Cucumber	1.06	–	–	–	2.70	26.70	–	–
		6	0.315	Tomato	28.00	–	–	–	2.50	20.50	–	–
		7	1.223	Maize	1.94	–	–	–	2.60	19.40	–	–

(Continued)

Table 2.6 (Continued)

Watershed	Farm	Plot	Area	Land use	Bray P	K	pH–H_2O	Clay + silt	Total soil N	Total soil C	Total tree C stock	Total tree C stock
			ha		mg kg^{-1}	cmol$_c$ kg^{-1}		%	g N kg^{-1}	g C kg^{-1}	kg C	kg C ha^{-1}
Sagana	SG1	1	0.0022	Arrow root	6.28	1.62	5.68	75.50	3.40	42.10	410	1,237
		2	0.063	Sweet potato	3.70	0.59	5.14	–	2.20	24.50	–	–
		4	0.18	Grass	1.87	0.26	5.28	–	2.60	25.50	–	–
		3	0.039	Napier grass	1.19	–	–	–	2.40	21.00	–	–
		5	0.11	Bean	1.67	–	–	–	2.70	24.40	–	–
		6	0.028	Fallow	1.64	–	–	–	2.40	–	–	–
		7	0.027	Grass	2.09	–	–	–	3.00	24.60	–	–
		8	0.094	Grass	5.03	–	–	–	3.10	31.20	–	–
		9	0.016	Arrow root	7.95	–	–	–	2.90	28.90	–	–
		10	0.02	Arrow root	2.74	–	–	–	2.90	18.00	–	–
		11	0.15	Tea	8.02	0.65	4.33	–	3.60	36.40	–	–
		12	0.021	Fallow	7.85	–	–	–	2.70	–	–	–
Sagana	SG2	1	0.033	Maize	3.04	0.55	5.03	85.70	2.60	25.20	11,079	52,258
		2	0.012	Napier grass	1.36	0.17	4.64	–	2.60	17.80	–	–
		3	0.261	Tea	16.40	0.66	4.62	–	2.60	22.60	–	–
		4	0.017	Potato (Irish)	9.51	–	–	–	2.30	19.40	–	–
		5	0.018	Napier grass	21.20	–	–	–	2.60	21.50	–	–
		6	0.185	Maize	35.50	–	–	–	2.20	20.50	–	–

Table 2.7 Farm area (m^2) and percentage of total farm area under various levels of slope steepness

Slope category, %	0–5		5–10		10–20		20–30		30+	
Area	m^2	%	m^2	%	m^2	%	m^2	%	m^2	%
Farm TK1	98.00	8.00	73.00	6.00	66.00	5.00	–	–	1,083.00	81.00
Farm TK2	42.00	0.00	952.00	7.00	112.00	1.00	2,078.00	15.00	10,240.00	76.00
Farm MR1	2,196.00	50.00	–	–	1,557.00	35.00	–	–	676.00	15.00
Farm MR2	–	–	269.00	4.00	1,846.00	25.00	5,127.00	70.00	61.00	1.00
Farm SG1	–	–	983.00	32.00	256.00	8.00	1,492.00	48.00	379.00	12.00
Farm SG2	1,869.00	88.00	–	–	72.00	3.00	48.00	2.00	136.00	6.00

Table 2.8 Estimation of soil loss per year per plot and per farming system

Plot no.[a]	Cover	Slope		Plot area	Soil loss[b]	SLM/crop erosion reduction factor assumed per plot	Reason	Adjusted soil loss	Soil loss	Total soil loss	Total soil loss[c]
		degrees	%	m^2	$Mg\,ha^{-1}$			$Mg\,ha^{-1}$	$Mg\,plot^{-1}$	$Mg\,farm^{-1}$	avg. $Mg\,ha^{-1}$
TK1—traditional farm with dairy production						–	–	–	–	–	–
12	Home garden	2.5	4.37	7.20	0.00	–	–	0.00	0.00	–	–
8	Maize/banana	2.6	4.54	90.74	0.00	0.20	Crop	0.00	0.00	–	–
4	Maize	4.8	8.40	41.37	159.10	–	–	159.10	0.66	–	–
1	Vegetables/potato	5.2	9.10	31.72	159.10	–	–	159.10	0.50	–	–
3	Bean	9.2	16.20	48.31	156.70	–	–	156.70	0.76	–	–
2	Napier grass	10.1	17.81	17.46	156.70	0.80	Crop	31.34	0.05	–	–
9	Fallow/potato	29.1	55.66	56.98	146.40	–	–	146.40	0.83	–	–
10	Kale/potato	29.5	56.58	55.41	146.40	–	–	146.40	0.81	–	–
5	Maize	35.4	71.07	156.48	146.40	0.90	Fanya juu	14.64	0.23	–	–
6	Pineapple/avocado	38.5	79.54	601.41	146.40	0.90	Fanya juu	14.64	0.88	–	–
7	Napier grass	34.7	69.24	167.46	29.28	0.80	Crop	5.86	0.10	4.83	37.88
TK2—commercially oriented cabbage, tea, and livestock						–	–	–	–	–	–
12	Arrow root	1.2	2.09	42.38	0.00	–	–	0.00	0.00	–	–
7	Fallow/maize	3.2	5.59	550.72	33.30	–	–	33.30	1.83	–	–
6	Cabbage	3.5	6.12	190.37	33.30	–	–	33.30	0.63	–	–
8	Maize/cabbage	3.7	6.47	210.96	33.30	–	–	33.30	0.70	–	–
5	Potato/maize/cabbage	10.1	17.81	112.45	159.10	–	–	159.10	1.79	–	–
9	Potato/maize	11.8	20.89	694.42	156.70	–	–	156.70	10.88	–	–
2	Maize/cabbage	13.3	23.64	143.13	156.70	–	–	156.70	2.24	–	–
1	Napier grass	13.6	24.19	1,240.80	156.70	0.80	Crop	31.34	3.89	–	–
3	Tea	18.1	32.69	3,254.49	14.64	0.90	Crop	1.46	0.48	–	–

Plot no.[a]	Cover	Slope		Plot area	Soil loss[b]	SLM/crop erosion reduction factor assumed per plot	Reason	Adjusted soil loss	Soil loss	Total soil loss	Total soil loss[c]
		degrees	%	m^2	$Mg\,ha^{-1}$			$Mg\,ha^{-1}$	$Mg\,plot^{-1}$	$Mg\,farm^{-1}$	avg. $Mg\,ha^{-1}$
11	Maize/cabbage	18.4	33.27	398.25	146.40	–	–	146.40	5.83	–	–
10	Tea	19.7	35.81	5,247.40	146.40	0.90	Crop	14.64	7.68	–	–
4	Napier grass	26.2	49.21	1,339.70	146.40	0.80	Crop	29.28	3.92	39.88	29.71
MR1—traditional farm with dry season cash crops						–	–	–	–	–	–
3	Napier grass	0.8	1.40	13.64	0.00	0.80	Crop	0.00	0.00	–	–
1	Maize	1	1.75	977.41	0.00	–	–	0.00	0.00	–	–
2	Napier grass	1.2	2.09	53.26	0.00	0.80	Crop	0.00	0.00	–	–
9	Maize	2.2	3.84	1,151.80	0.00	–	–	0.00	0.00	–	–
11	Maize and bean	6.1	10.69	823.35	159.10	0.90	Fanya juu	15.91	1.31	–	–
4	Maize and bean	7.4	12.99	87.15	159.10	0.90	Fanya juu	15.91	0.14	–	–
10	Maize and bean	7.9	13.88	646.41	159.10	0.90	Fanya juu	15.91	1.03	–	–
6	Napier grass	18.6	33.65	36.87	146.40	0.97	Terraces	4.39	0.02	–	–
5	Maize and bean	22.7	41.83	14.07	146.40	0.97	Terraces	4.39	0.01	–	–
8	Maize	23.5	43.48	17.93	146.40	0.97	Terraces	4.39	0.01	–	–
7	Banana	23.3	43.07	606.71	146.40	0.20	Crop	117.12	7.11	9.61	21.71
MR2—commercially oriented tomato farm						–	–	–	–	–	–
2	Potato/maize/tomato	3.6	6.29	237.23	33.30	0.97	Terraces	1.00	0.02	–	–
4	Bean	4.8	8.40	32.09	33.30	0.97	Terraces	1.00	0.00	–	–
1	Maize/bean/tomato	5.9	10.33	570.42	159.10	0.97	Terraces	4.77	0.27	–	–
6	Tomato/maize	7.1	12.46	1,275.50	159.10	0.97	Terraces	4.77	0.61	–	–
7	Tomato/maize	11.7	20.71	4,951.20	156.70	0.97	Terraces	4.70	2.33	–	–
3	Maize/bean/tomato	11.8	20.89	175.80	156.70	0.97	Terraces	4.70	0.08	–	–
5	Cucumber/fallow	31.5	61.28	61.26	146.40	0.97	Terraces	4.39	0.03	3.35	4.58

(Continued)

Table 2.8 (Continued)

Plot no.[a]	Cover	Slope		Plot area	Soil loss[b]	SLM/crop erosion reduction factor assumed per plot	Reason	Adjusted soil loss	Soil loss	Total soil loss	Total soil loss[c]
		degrees	%	m^2	$Mg\,ha^{-1}$			$Mg\,ha^{-1}$	$Mg\,plot^{-1}$	$Mg\,farm^{-1}$	avg. $Mg\,ha^{-1}$
SG1—traditional farm with some horticultural cash crops and tea						–	–	–	–	–	–
4	Pasture	4.7	8.22	729.35	33.30	0.80	Crop	6.66	0.49	–	–
2	Sweet potato	4.9	8.57	253.43	33.30	–	–	33.30	0.84	–	–
10	Maize/bean	6.1	10.69	80.43	159.10	–	–	159.10	1.28	–	–
1	Arrow root	9.6	16.91	89.33	159.10	–	–	159.10	1.42	–	–
12	Fallow	9.9	17.45	86.61	159.10	–	–	159.10	1.38	–	–
3	Napier grass	13.7	24.38	155.89	156.70	0.97	Terraces	4.70	0.07	–	–
5	Maize/bean/potato	14.8	26.42	443.43	156.70	0.97	Terraces	4.70	0.21	–	–
6	Fallow/sweet potato	15	26.79	113.40	156.70	0.97	Terraces	4.70	0.05	–	–
11	Tea	16.3	29.24	605.05	15.67	0.90	Crop	1.57	0.09	–	–
7	Pasture	16.5	29.62	109.65	31.34	0.97	Terraces	0.94	0.01	–	–
9	Maize/bean/ vegetables	15.4	27.54	64.57	156.70	–	–	156.70	1.01	–	–
8	Pasture	17.8	32.11	378.99	29.28	0.80	Crop	5.86	0.22	7.08	22.77
SG2—traditional farm with coffee						–	–	–	–	–	–
2	Coffee	0.2	0.35	1,054.80	0.00	0.60	Crop	0.00	0.00	–	–
4	Potato/Napier grass	0.2	0.35	67.17	0.00	0.80	Crop	0.00	0.00	–	–
6	Maize/bean/cabbage	1.9	3.32	746.86	0.00	0.97	Terraces	0.00	0.00	–	–
5	Napier grass	7.1	12.46	72.06	31.82	0.97	Terraces	0.95	0.01	–	–
3	Napier grass	12.2	21.62	47.88	31.34	0.80	Crop	6.27	0.03	–	–
1	Maize/bean	22.1	40.61	135.73	146.40	0.97	Terraces	4.39	0.06	0.10	0.45

[a] From least to most sloped farm.
[b] Based on Mutegi et al. (2008).
[c] This column gives the values that were used and reported on in this study.

These results confirm the high risk for soil loss in the Central Highlands of Kenya. Mutegi et al. (2008) measured soil loss between 33 and 160 Mg soil ha^{-1} yr^{-1}, and Angima et al. (2003) report up to 150–200 Mg ha^{-1} yr^{-1}. This represents a substantial loss of both soil nutrients and OM. Although farms differed in the steepness of their plots, all six farms had in place some practices for soil erosion control, mainly in the form of terraces and fanya juu (Table 2.8), which highly reduced soil loss. However, not all soil protection measures were in a good state, and farmers and extension officers expressed the need to maintain and further expand these practices.

The N balances were positive on all farms, indicating an overall accumulation and even risk of environmental loss of N, especially in the Thika–Chania farms (Figure 2.3). Farm TK1 (traditional farm with dairy production), had the highest N balance (178 kg N ha^{-1}), which was primarily due its high stocking rate. It was the smallest sized farm but kept two adult cows and a calf for which large amounts of N per area had to be imported through purchased hay, concentrates, and cut-and-carry grasses from roadsides. Nitrogen is lost from manure due to poor management and storage practices. As crop production on this farm was focused on household consumption, N exports through sale of crop products was negligible. Farm SG1 (traditional farm with some horticultural cash crops and tea) had the third highest N balance (77 kg N ha^{-1}), also due to imported feeds through grazing cattle on external grassland. On TK2 (the commercially oriented cabbage, tea and livestock farm), the positive N balance (119 kg N ha^{-1}) could also be largely contributed to the import of feeds, especially poultry feed for the 900 chickens and dairy meal. Farm MR1 (N balance 58 kg N ha^{-1}) (traditional farm with dry season cash crops) and MR2 (44 kg N ha^{-1}) (the commercial tomato farm) had the highest contributions of mineral fertilizer to their N inputs, which was applied to maize, dry

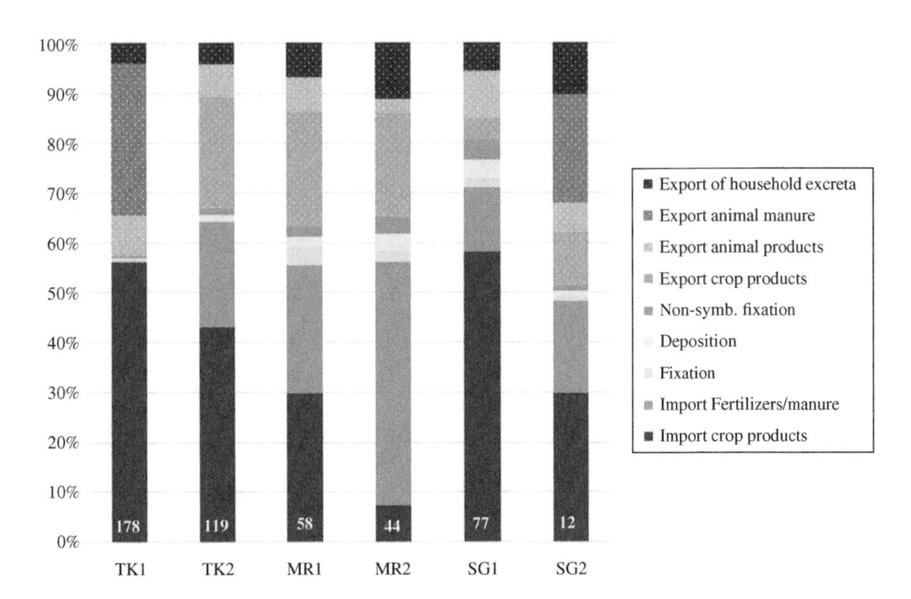

Figure 2.3 Inputs (bold colors) and outputs (patterns) for the farm N balance per farm, with total N balance amounts indicated in white numbers at the bottom of the bars.

season vegetables, and tomatoes. Although all farms exported some crop and livestock products, legume cropping contributed very little to the overall N balance. Farm SG2 (the traditional farm with coffee) had the lowest N balance. Nitrogen imports through dairy meal, cut-and-carry grasses, and mineral fertilizers used for coffee, maize, bean, cabbage, and Napier grass were balanced with N losses from manure and excreta and crop and livestock product sales.

GHG Emissions

Greenhouse gas (GHG) emissions were calculated per farm, per area, and per unit milk production and converted into Mg CO_2 equivalent, with Figure 2.4 showing per area emissions. Livestock was the largest contributor to total GHG emissions across all farms, confirming findings from other studies in smallholder farming systems in East Africa (Ortiz–Gonzalo et al., 2017; Paul et al., 2018, 2020; Seebauer, 2014). Livestock contributes to the GHG balance through methane from enteric fermentation, which was the largest source on these farms, and nitrous oxide from manure storage and application. Highest GHG emissions per unit land area were calculated with the farms with the highest livestock rates (TK1, TK2, and SG2). Although farms were using mineral fertilizer on mainly tea and vegetables, nitrous oxide emissions from fertilizers were minor when compared with other emission sources. Unlike the other farms, emissions from manure on TK2 were almost half of those from enteric fermentation; this is due to the large poultry holdings, which do not emit methane. Emissions from the Upper Tana were higher when compared with other crop–livestock systems in Tanzania, which ranged from 1.14 to 2.6 Mg $CO_2e\,ha^{-1}$ (Paul et al., 2020), going back to higher production intensity in the Central Highlands of Kenya. Greenhouse gas emissions per unit milk production ranged from 0.962 to 1.508 kg $CO_2e\ kg^{-1}$, which is similar to what

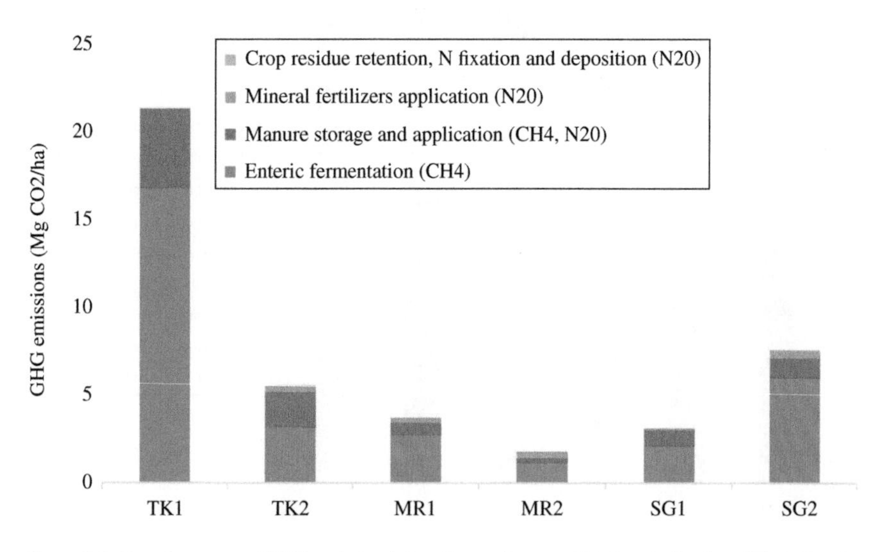

Figure 2.4 Greenhouse gas (GHG) emission intensity per farm and hectare indicating different emission sources.

Table 2.9 Operating profit and gross margins per farm in US$ yr^{-1}

Farm	TK1	TK2	MR1	MR2	SG1	SG2
Operating profit, $ yr^{-1}	642	5,588	1,061	2,951	312	599
Gross margin crops, $ yr^{-1}	133	5,208	585	2,604	153	332
Gross margin animals, $ yr^{-1}	509	526	526	485	168	305

Ortiz-Gonzalo et al. (2017) found for the same region (0.72–1.37 kg CO_2e kg^{-1} milk) but lower than for more extensive crop–livestock systems in northern Tanzania (2.15–15.32 kg CO_2e kg^{-1} milk, Paul et al., 2020).

Profitability

The two more commercially oriented farms, TK2 and MR2, were also the most profitable with US$5,588 yr^{-1} and $2,951 yr^{-1} operating profit. As both were cash crop oriented, the gross margin for crops contributed almost entirely to profits (Table 2.9). All farms were earning income from milk production in addition to cash crops and a surplus of staple crops. The smaller, more traditional farms (TK1 and SG2) had one to two dairy cows. For these farms, the sale of milk contributed substantially to profits compared with their fewer staple and cash crops.

Scenario Impacts

The maintenance scenario did not have an impact on the N balance across farms (Figure 2.5a–2.5f). The grass strip scenarios had some impacts on the N balance at farm level, both reducing or increasing depending on the particular farm setup and practices. In the case of TK1 (Figure 2.5a), grass strips reduced N balance of up to 8 kg N ha^{-1}, as they absorbed some N in biomass growth and increased milk production that was sold from the farm. In the case of MR1 (Figure 2.5c) and MR2 (Figure 2.5d), the implementation of Napier strips decreased the need to import feed and mineral fertilizer because the grass strips occupied some of the land previously cultivated with cash crops, resulting in an overall decrease of the N balance.

All three scenarios had mostly a negative impact on the profitability of the farms (Figure 2.5), although these only ranged from −0.10% (Brachiaria strips in TK2) and −0.55% (Brachiaria strips in SG2) to −9.26% (maintenance in SG1). In the case of the maintenance scenario, the additional costs derived from labor required for the upkeep of the erosion control infrastructures. In the case of the grass strip scenarios, the assumed increase in milk production by 5% for Brachiaria balanced out some of the loss in revenue from the decrease in crop area. Only in the case of TK1 (Figure 2.5a) does this increase in profits from milk sales can outweigh the loss in revenue from lost cropping area.

The impact of the scenarios on GHG emission intensity per unit milk produced can be neutral, positive, or negative, depending on the particular setup and practices of the farms. Maintenance did not impact GHG emissions, as it implied no changes for livestock feeding and productivity or fertilizer application. In farms with high fertilizer application, reducing the area of crops to plant grass strips may reduce associated GHG emissions. Grass strips, and especially Brachiaria strips, improved feed quality provided to cattle and increased milk production and could therefore reduce methane emissions from enteric fermentation. There

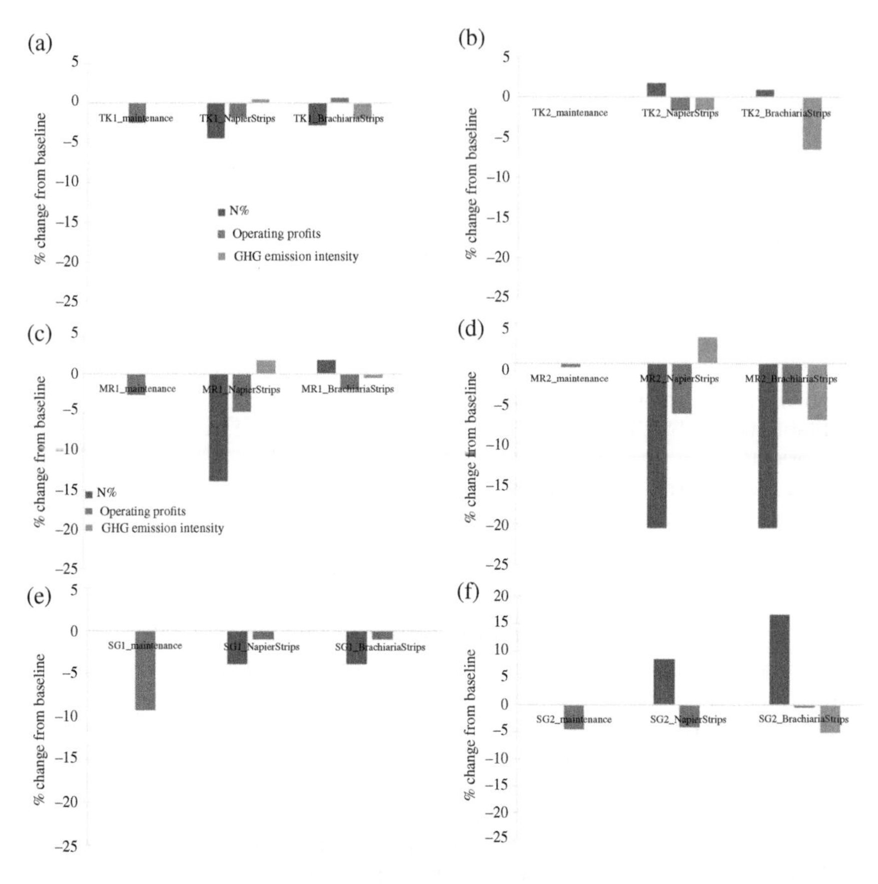

Figure 2.5 Changes through scenarios "maintenance," "NapierStrips," and "BrachiariaStrips" in outcome indicators N balance, operating profits, and greenhouse gas (GHG) emissions across all farms (a) TK1, (b) TK2, (c) MR1, (d) MR2, (e) SG1, and (f) SG2.

was a reduction from 0.005 to 0.1 kg CO_2e kg^{-1} of milk (Figure 2.5b, d, f) on TK2, MR2, and SG2 under the Brachiaria scenario. Increase in feed through Napier grass did not result in increased milk production but did increase N excretion and therefore loss through manure, increasing GHG intensity per unit milk produced on TK1, MR1, and MR2 (Figure 2.5a, c, d).

Conclusions

This study offers valuable insights for the Nairobi Water Fund and decision makers and designers of similar interventions and areas. First, it illustrates the heterogeneity of the farms in the three sub-watersheds in terms of types of soils, crops, livestock, and their management practices. This heterogeneity impacts the soil erosion reduction that can be achieved, but also potential co-benefits and trade-offs in terms of N balance, profitability, and GHG emissions. Second, profitability trade-offs for SLM practices exist, especially in cases where higher labor investment (e.g., for terrace maintenance) is not compensated for by other income sources (e.g., higher milk production from improved feed obtained from grass strips). These are likely to be strong determinants for SLM adoption, with SLM practices with co-benefits having a higher likelihood of finding wider acceptance. Thirdly, grass strips in integrated crop–livestock systems can be a key entry point for sustainable development programs, reducing erosion and providing associated co-benefits such as reduced GHG emission intensity. Livestock should be a key part of development programs focused on natural resource management, rather than avoided. Lastly, this study demonstrates the added value to linking bio-economic modeling and participatory socio-economic assessments for contextualized ex-ante impact assessments to inform development interventions.

Acknowledgments

This study was conducted under the project "Water, Land and Ecosystems in Africa: Restoring Degraded Landscapes through Selective Investments in Soil Quality," funded by the International Fund for Agricultural Development (IFAD) and managed by the International Water Management Institute (IWMI) (Agreement reference 4500022483). The work was undertaken as part of the CGIAR Research Program on Land, Water and Ecosystems (WLE), which is carried out with support from the CGIAR Fund Donors and bilateral funding agreements. We appreciate the help of Jane Gicheha and Jessica Koge from CIAT, who assisted in household data collection, and John Mutua from CIAT for plotting Figure 2.2.

Abbreviations

AGB, aboveground biomass; DBH, diameter at breast height; DM, dry matter; ELMO, Evaluating Land Management Options; GHG, greenhouse gas; GPS, global positioning system; OM, organic matter; SLM, sustainable land management; TNC, The Nature Conservancy.

References

Angima, S., Stott, D. E., O'Neill, M. K., Ong, C. K., & Weesies, G. A. (2003). Soil erosion prediction using RUSLE for central Kenyan highland conditions. *Agriculture, Ecosystems and Environment, 97*, 295–308.

Braslow, J., Cordingley, J. (2016). *Participatory mapping in the Upper Tana River Basin, Kenya. A case study.* International Center for Tropical Agriculture (CIAT).

Groot, J. C. J., Oomen, G. J. M., & Rossing, W. A. H. (2012). Multi-objective optimization and design of farming systems. *Agricultural Systems, 110*, 63–77.

Kuyah, S., Dietz, J., Muthuri, C., Jamnadass, R., Mwangi, P., Coe, R., & Neufeldt, H. (2012). Allometric equations for estimating biomass in agricultural landscapes: I. Aboveground biomass. *Agriculture, Ecosystems & Environment, 158*, 216–224.

Muinga, R. W., Njunie, M. N., Gatheru, M., & Njarui, D. M. G. (2016). The effects of Brachiaria grass cultivars on lactation performance of dairy cattle in Kenya. In D. M. G. Njarui, E. M. Gichangi, S. R. Ghimire, & R. W. Muinga (Eds.), *Climate-smart Brachiaria grasses for improving livestock production in East Africa: Kenya experience. Proceedings of the Workshop, Naivasha, Kenya, 14–15 September* (pp. 229–237). Kenya Agricultural and Livestock Research Organization.

Mutegi, J. K., Mugendi, D. N., Verchot, L. V., & Kung'u, J. B. (2008). Combining Napier grass with leguminous shrubs in contour hedgerows controls soil erosion without competing with crops. *Agroforestry Systems, 74*, 37–49. https://doi.org/10.1007/s10457-008-9152-3

Ortiz-Gonzalo, D., Vaast, P., Oelofse, M., Neergaard, d. N.,. A., Albrecht, A., & Rosenstock, T. S. (2017). Farm-scale greenhouse gas balances, hotspots and uncertainties in smallholder crop–livestock systems in Central Kenya. *Agriculture, Ecosystems, and Environment*, *248*, 58–70. https://doi.org/10.1016/j.agee.2017.06.002

Paul, B. K., Birnholz, C., Timler, C., Michalscheck, M., Koge, J., Groot, J., & Sommer, R. (2015). *Assessing and improving organic matter, nutrient dynamics and profitability of smallholder farms in Ethiopia and Kenya: Proof of concept using the whole farm model FarmDESIGN for trade-off analysis and prioritization of GIZ development interventions*. CIAT Working Document 408, International Center for Tropical Agriculture. https://cgspace.cgiar.org/handle/10568/69149?show=full

Paul, B. K., Frelat, R., Birnholz, C., Ebong, C., Gahigi, A., Groot, J. G. J., Herrero, M., Kagabo, D., Notenbaert, A., Vanlauwe, B., & van Wijk, M. T. (2018). Agricultural intensification scenarios, household food availability and greenhouse gas emissions in Rwanda: Ex-ante impacts and trade-offs. *Agricultural Systems*, *163*, 16–26. https://doi.org/10.1016/j.agsy.2017.02.007

Paul, B. K., Groot, J. C. J., Birnholz, C. A., Nzogela, B., Notenbaert, A., Woyessa, K., Sommer, R., Nijbroek, R., & Tittonell, P. (2020). Reducing agro-environmental trade-offs through sustainable livestock intensification across smallholder systems in northern Tanzania. *International Journal of Agricultural Sustainability*, *18*(1), 35–54. https://doi.org/10.1080/14735903.2019.1695348

Rufino, M. C., Quiros, C., Boureima, M., Desta, S., Douxchamps, S., Herrero, M., Wanyama, I. (2012). *Developing generic tools for characterizing agricultural systems for climate and global change studies (IMPACTlite-phase 2)*. CGIAR Research program on Climate Change, Agriculture and Food Security (CCAFS). https://hdl.handle.net/10568/42065

Seebauer, M. (2014). Whole farm quantification of GHG emissions within smallholder farms in developing countries. *Environmental Research Letters*, *9*(3), 1–19. https://doi.org/10.1088/1748-9326/9/3/035006

Timler, C., Alvarez, S., DeClerck, F., Remans, R., Raneri, J., Estrada Carmona, N., Mashingaidze, N., Chatterjee, S. A., Chiang, T. W., Termote, C., Yang, R. Y., Descheemaeker, K., Brouwer, I. D., Kennedy, G., Tittonell, P. A., & Groot, J. C. J. (2020). Exploring solution spaces for nutrition-sensitive agriculture in Kenya and Vietnam. *Agricultural Systems*, *180*, 102774. https://doi.org/10.1016/j.agsy.2019.102774

TNC (The Nature Conservancy). (2015). *Upper Tana–Nairobi Water Fund Business Case. Version 2*. The Nature Conservancy. https://www.nature.org/content/dam/tnc/nature/en/documents/Nairobi-Water-Fund-Business-Case_FINAL.pdf

3

Aleme Araya, P. V. Vara Prasad,
Ignacio A. Ciampitti,
Charles W. Rice, and
Prasanna H. Gowda

Using Crop Simulation Models as Tools to Quantify Effects of Crop Management Practices and Climate Change Scenarios on Wheat Yields in Northern Ethiopia

Abstract

A simulation model Decision Support System for Agrotechnology–Cropping Systems Model (DSSAT–CSM) was used to identify optimum level of agronomic practices for growing wheat (*Triticum aestivum* L.) in sandy loam soils in northern Ethiopia, and to assess impacts of climate change and management strategies on wheat yields. The model was calibrated for its soil and cultivar parameters at the site in an earlier study. The simulated practices included seven planting dates (each separated by 10 d starting from 20 June to 20 August), three plant densities (100, 200, and 300 plants m^{-2}), two irrigation treatments (two irrigations at time of flowering [simulating water savings in tied ridges] and dryland), and 13 N fertilizer rates (at increments of 16 kg N ha^{-1} from 0 to 192 kg ha). The 30-yr (1980–2009) baseline climate was obtained from the study site whereas the future climate was generated based on two global climate models (HadGEM2-ES and GFDL-ESM2G). The wheat crop was exposed to different combinations of agronomic practices and climate inputs. Results showed that early planting resulted in highest yield whereas later planting substantially decreased yield. Higher plant density of 300 plants m^{-2} resulted in greater yield under different planting dates and fertilizer rates. Irrigation

Enhancing Agricultural Research and Precision Management for Subsistence Farming by Integrating System Models with Experiments, First Edition. Edited by Dennis J. Timlin and Saseendran S. Anapalli.

DOI: 10.1002/9780891183891.ch03

treatments resulted in higher yields under all conditions relative to dryland; however, the responses to irrigation were greater for later planting dates. Yield increase ranged from 8 to 25% for every increment of fertilizer N applied as 16 kg N ha^{-1}. The response of wheat to future mid-century (2040–2069) climate scenario showed no negative impact on yield compared with baseline climate (1980–2009) under similar management practices. This suggested that the increase in temperature in the mid-century stayed within the optimum range for the wheat cultivar, negative effect of increased temperature was compensated by the beneficial effect of increased CO_2, and the change in rainfall was not significant. Overall, this exercise shows how crop models can be used for identifying suitable management practices and to forecast potential impacts of climate change on crop yields. Such information will be useful to agronomists, plant biologists, and policy makers for developing strategies for increasing or sustaining crop yields under current and future changing environments and management practices.

Introduction

In Ethiopia, cereals are the major food crops, occupying more than 70% of the cultivable land area and contributes to 14% of the total gross domestic product (Taffesse et al., 2012). Wheat (*Triticum aestivum* L.) is one of the major cereal crops grown in Ethiopia. It is an important source of food for humans and animals. Wheat accounts for about 13% of all crops and is ranked between third and fourth among cereals cultivated in Ethiopia (Taffesse et al., 2012; CSA, 2000–2018). There has been an increase in wheat production during the past decade attributed to an increase in cultivated land (Taffesse et al., 2012) and inputs (CSA, 2000–2018), but the yield is still far below the attainable yield due to several abiotic, biotic, and socio-economic factors. Some of the key abiotic factors include land degradation (Stroosnijder, 2009), water stress, poor soil fertility, and nutrient deficiencies (Araya et al., 2019a; Silungwe et al., 2018; Stewart et al., 2020). The biotic factors include the occurrence of pests, diseases, and ineffective weed control practices. The socio-economic factors include the lack of availability and high cost of improved seeds and inorganic sources of fertilizers, limited access to market and stable market prices, and unavailability of credit. Improving yields will require addressing these key issues to ensure that there is an enabling environment for production of wheat and other cereal grain crops.

Improved crop, soil, water, and nutrient management practices are critical to sustain and enhance crop productivity. Studies have indicated that the use of combinations of best agronomic management practices such as optimal planting date, plant density, fertilizer, and improved water management practices—such as tied ridging and irrigation—could enhance crop yield and contribute to food security (Araya et al., 2019a; Majaivana et al., 2016). Numerous best practices have been developed through conventional field experiments that are expensive, time-consuming, and are often difficult to replicate over a range of agro-ecological conditions. Improved technologies must be tested at wider spatial- and temporal-scale before promoting adoption because of the diverse nature of soils, climate, crops, cultivar characteristics, and socio-economic conditions of the farmers. With the advancement in mechanistic crop simulation modeling and analytical tools for large-scale data, there are new opportunities to use such models for capturing impacts of cultivar, crop management (e.g., crop, water, and nutrients), and climate change factors (Araya et al., 2019a, 2019b, 2019c, 2020a, 2020b; Boote et al., 2012; Singh et al., 2015). Crop simulation models do not replace traditional experiments, but rather complement them. Selection, calibration, and validation of the crop models is important before using them to evaluate a wide range of management and climate scenarios. Crop models also need to be accurately parametrized to ensure that they are tested for the impacts of climate factors on growth and development (Boote et al., 2011, 2018). Once calibrated and validated, the crop models can assist to evaluate different management strategies across time and space by considering soils, climate, and crop characteristics (Araya et al., 2017a, 2017b, 2017c, 2020a, 2020b; Boote et al., 2012; Silungwe et al., 2018). These tools are useful and improve efficiency, reduce time and cost, and increase wide representation of geographies (Boote et al., 2012). There is also a continuous need to design crop science experiments to inform and refine responses of crop models to management and climate change factors (Boote et al., 2005, 2018; Craufurd et al., 2013).

The Decision Support System for Agrotechnology–Cropping Systems Model (DSSAT–CSM; Jones et al., 2003; Hoogenboom et al., 2015) has been applied for assessing impacts of climate change on various crops in different geographical locations (Araya et al., 2015, 2020a, 2020b; Jones & Thornton, 2003; Joshi et al., 2015; Ruane et al., 2013). Araya et al. (2015) utilized both DSSAT and APSIM (Agricultural Production System sIMulator) models for exploring impacts of climate change on maize in southwestern Ethiopia, portraying that climate change may not cause significant yield reduction during the mid-century period. This is consistent with results reported in the literature (Jones & Thornton, 2003). As such, information on potential impacts of climate change and adaptation strategies are also important for Ethiopian wheat and other food crops (Araya et al., 2019a, 2020a, 2020b), as they are greatly influenced by climate under dryland conditions, especially by the climate in northern Ethiopia as it is highly variable with a short growing season (Araya et al., 2012).

Agriculture in northern Ethiopia is highly dependent on rains and subject to frequent and prolonged droughts. Consequently, food insecurity exists in this region. It is possible that many of those regions currently affected by climate change and variability could be further challenged by increased food insecurity under future climates. By the end of this century, global temperature is expected to increase by 2.6–4.8 °C, under Representative Concentration Pathways (RCP) of 8.5, relative to the 1966–2005 period (IPCC, 2014). A rise in air temperature by 1 °C could reduce global wheat production by 6% (Asseng et al., 2011, 2015; Liu et al., 2016). Literature on yield loss due to high temperature stress during reproductive stages of crop development are available (Prasad et al., 2006, 2015, 2017). High temperature stress decreases seed-set percentage (Barlow et al., 2015; Nuttall et al., 2017; Prasad & Djanaguiraman, 2014; Prasad et al., 2015, 2017), consequently decreasing grain yield due to lower grain numbers and smaller grain size as a result of shorter grain-filling duration (Prasad et al., 2015, 2017).

Crop models are an efficient way to investigate and identify suitable agronomic management practices, both under present and future scenarios for improving and sustaining wheat productivity. In this study, the DSSAT–CSM was used to simulate impacts of different crop, water, and nutrient management practices under baseline and future climate scenarios on wheat yield in northern Ethiopia. The objectives of this research were to (a) identify suitable agronomic practices for improving wheat yields in sandy loam soils and (b) assess impacts of climate change and management strategies on wheat yields.

Materials and Methods

Study Site

The study site was located at Adigudom, (13°15′ E and 39°30′ N, with an elevation of 2,068 m asl) in northern Ethiopia. The average mean annual rainfall was 643 mm, and daily minimum and maximum temperature for Adigudom was 26.7 and 13.8 °C, respectively. This is one of the major wheat growing locations that represent the vast majority of northern Ethiopia in terms of climate, cropping pattern, crop management, agro-ecology, and soil. The study was carried out on sandy clay loam soils with a permanent wilting point, field capacity, and water content at saturation of 0.15–0.18, 0.26–0.29, and 0.38–0.43 m^3 m^{-3}, respectively, with bulk density of about 1.41–1.47 g cm^{-3}.

Model Setup and Agronomic Management Simulation Scenarios

The DSSAT–CSM was calibrated and validated for wheat variety (HAR-2501) in a prior study (Araya et al., 2019a). The following initial assumptions were set in the model: crop residue cover was approximately 1,000 kg ha^{-1}; previous crop was assumed to be wheat as continuous wheat is commonly practiced; and soil water (top 0.3 m) at planting was assumed to be 60% of the upper drainage limit. This initial assumption of soil water was based on field measurements at nearby locations where the model was calibrated (Araya et al., 2019a). However, this assumes normal onset of rains. Thus, it is important to note that the planting date treatments presented in this study are impacted by the initial soil water assumption.

A prior study verified the strong relationship between simulated and measured yield for the calibration datasets (R^2 = .87) (Araya et al., 2019a). The difference between the measured and simulated values were low with normalized root means square of error of 19% and with an index of agreement of .98. In addition, the model validation datasets showed a strong relationship between simulated and measured yield (R^2 = .94 and index of agreement of .74) with normalized root mean square of error of 10.2% (Araya et al., 2019a).

The wheat genotype (the genetic coefficients as calibrated in DSSAT–CSM based on prior studies) was simulated under various agronomic practices. The general plant date in the region is from mid-June to late August depending on the resource availability (labor and draft animals), onset of rain and crop and variety type. For this case, the variety HAR-2501 is an early maturing wheat variety. Planting for an early maturing type usually begins from early July and ends around mid-August. However, for scenario analysis, the simulated treatments included seven planting dates (20 June, 1 July, 10 July, 20 July, 1 August, 10 August, and 20 August); three plant densities (100, 200, and 300 plants m^{-2}), two water treatments (dryland or rainfed and two irrigation events scheduled on 15 and 20 September of each year [time when the rainfall ceased] with a depth of 30 mm with an efficiency of 0.5); and 13 fertilizer N rates (0, 16, 32, 48, 64, 80, 96, 112, 128, 144, 160, 186, and 192 kg ha^{-1}). The two irrigations (each 30 mm during the flowering stage) were included to assume impacts of tied ridging on yield of wheat (unpublished) as established based on experimental data obtained from earlier studies

(Araya et al., 2010). Graphs were prepared for visual comparison of yield against the different planting dates and yield vs. N fertilization rate for the different planting dates and plant densities. In addition, percentage change in response to agronomic management practices over the controls were presented.

Simulation Scenarios for Climate Change Risk Assessment

Climate data was generated based on two Global Climate Models (HadGEM2-ES [Jones et al., 2011] and GFDL-ESM2G [Dunne et al., 2012]) under relatively higher greenhouse gas concentration pathway (RCP8.5) for mid-century period (2040–2069) using an AgMIP climate generation tool (AgMIP, 2013). The generated climate data and estimated CO_2 (an average estimated CO_2 concentration for mid-century based on Rosenzweig et al. (2016), 571 μmol mol^{-1}) were used as input in the crop model. The average daily minimum and maximum air temperatures for the months in the year for the baseline (1980–2009) and future (1940–2069) period were 26.7 and 29.5 °C and 13.8 and 16.7 °C, respectively (Figure 3.1).

Effects of different agronomic management scenarios (treatments) presented in the sections above were compared both under mid-century (2040–2069) and baseline (1980–2009) climates. Particularly, the yield comparison was focused more on selected agronomic practices such as planting date (20 June–20 August) and fertilizer N rate (0–192 kg N ha^{-1}) for scenario analysis as the crop was very responsive to both practices. Simulated future yield change was compared with the baseline yield and analyzed based on the following equation:

$$[(FY - BY)/(BY)] \times 100 \tag{3.1}$$

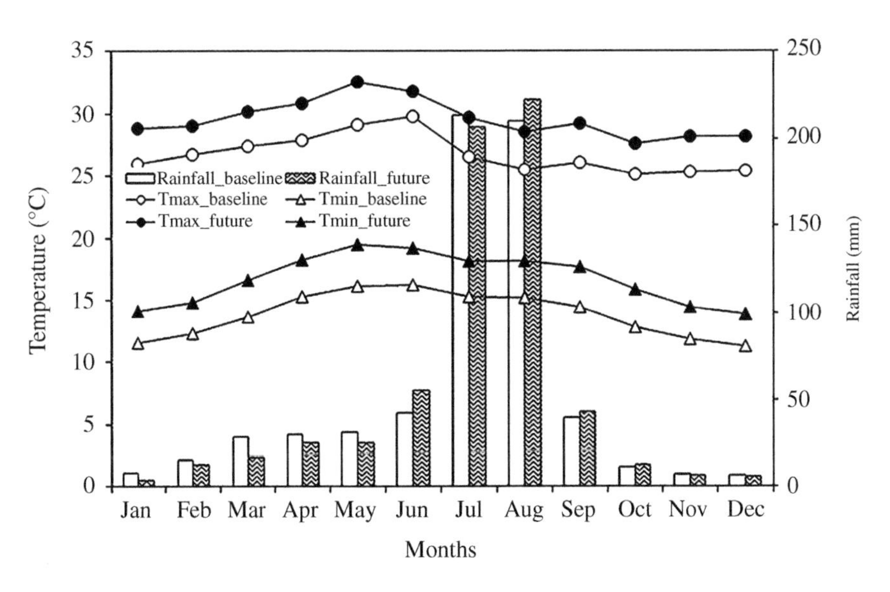

Figure 3.1 Average maximum and minimum temperatures for the months in a year during the baseline (1980–2009) and future (2040–2069) period (primary axis) and baseline and future monthly sum of rainfall (secondary axis) for Adigudom, northern Ethiopia. Tmin and Tmax, minimum and maximum temperatures, respectively.

where FY is future yield and BY is baseline yield [can be irrigation (as replacement for impacts of tied ridging) or dryland]. Baseline yield in this case is yield simulated based on the past 30 yr climate data (1980–2009), assuming a CO_2 concentration level of 380 μmol mol^{-1}.

Similarly, the average percentage yield increase per increase in N rate by 16 kg N ha^{-1} was estimated by comparing yield with the control for each planting date scenario Equation (3.2).

$$[(Y_n - Y_0)/(Y_0)] \times 100 \tag{3.2}$$

where Y_n is simulated wheat yield with fertilizer N application and Y_0 is simulated wheat yield without N fertilization (control) under each planting date. An average percentage yield increase due to an increase in N rate across all planting dates and fertilizers was also estimated using Equation (3.2) but with only one control (not varying by planting date) was used by averaging all controls (zero-N fertilizer) from all planting dates simulated under the dryland for the baseline scenario.

Regression analysis was used to examine the relationship between simulated future dryland yield and irrigated or dryland baseline yield to understand if climate changes caused yield irregularities due to the increase in temperatures when compared to the corresponding baseline yield. In addition, the relationship was used to develop regression equations so that others could use such empirical equations to generate future yield for similar locations in sandy loam soils.

Results and Discussion

Effects of Planting Dates

Early planting substantially improved wheat yield (Figure 3.2a, b). The highest (>4,500 kg ha^{-1}) and lowest (<2,000 kg ha^{-1}) yield was simulated for planting date 20 June and 20 August under planting density of 300 and 100 seeds m^{-2}, respectively (Figure 3.2). The results in Figure 3.2 are for the highest N fertilizer rate (192 kg ha^{-1}). Also, the highest yields of >4,500 kg ha^{-1} were obtained for the long-term average of the irrigated baseline, and yields were less for the dryland baseline. Many farmers in the study region plant wheat between 20 June and 20 August depending on the start of rainfall. Based on long-term climatic data, the model showed it was possible to optimize wheat yield by planting early (20 June). Araya et al. (2017a) conducted a simulation study and reported that planting wheat between 15 June and 5 July could improve yield. However, in some seasons, planting in June could be a bit risky due to false start of rains (Araya et al., 2012). However, at least planting around early July might minimize yield loss due to false start of rain. Araya and Stroosnijder (2011) reported about 50–89% of rain onset occur before 10 July and 44–46% of the rain cessation occur around 10 September. Long-term simulation studies by Araya et al. (2012) showed that the probability of crop failure due to false start for barley (*Hordeum vulgare* L.) planted on 3 July was approximately 20%; however, crop failure due to false start of rain slightly decreased to 17% for barley planted on 7 July. On the other hand, further delaying beyond 15 July may substantially reduce yield due to shortening of the rainy period during the growing season. In the current study, if planting was delayed by a month from 20 June to 20 July, wheat yield reduced substantially, mainly due to shortening of the growing period (Figure 3.2). Similarly,

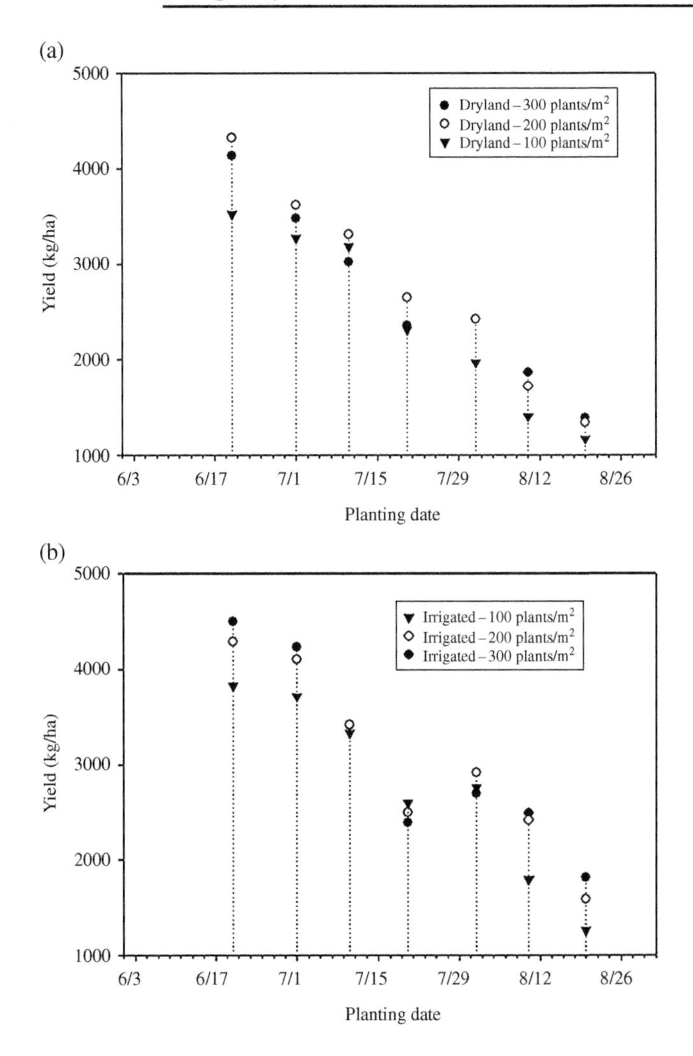

Figure 3.2 Effect of planting dates on wheat yield at varying plant density and water management (irrigation levels)—(a) dryland and (b) two irrigation = tied ridging practice—under high fertilizer N application rate (192 kg ha^{-1}) in sandy clay loam soils of Adigudom area, northern Ethiopia.

late planting (after 10 August) could expose the crop to late-season drought as rains in northern Ethiopia ceases around mid to late September (Araya & Stroosnijder, 2011; Araya et al., 2012). This shows crop models could help explore suitable planting strategies under various agro-ecologies and regions.

Effects of Plant Densities

There was a substantial difference in wheat yield between 300 and 100 plants m^{-2} for the selected planting dates (Figures 3.2 and 3.3). The model simulated the highest and lowest yields for plant densities of 300 and 100 plants m^{-2}, respectively.

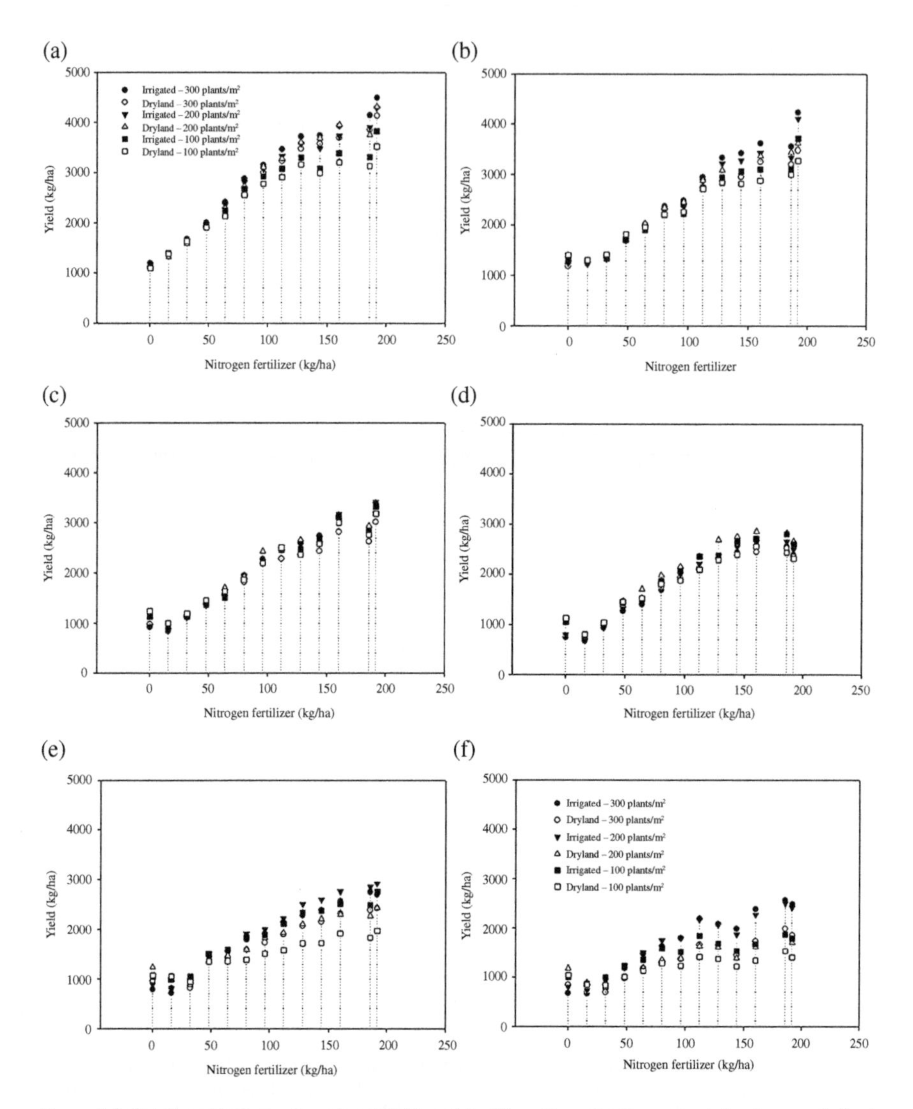

Figure 3.3 Relationship between wheat yield and fertilizer N application rate, irrigation, and plant density under different planting dates—(a) 20 June, (b) 1 July, (c) 10 July, (d) 20 July, (e) 1 August, and (f) 10 August—in sandy clay loam soils of Adigudom, northern Ethiopia.

The average yield difference between the two plant densities (300 and 100 plants m^{-2}) under different planting dates and fertilizer rates ranged from 500 to 1,000 kg ha^{-1}. Majaivana et al. (2016) conducted field experiments evaluating optimal plant densities and fertilizer applications for some selected wheat genotypes around the study site. They obtained high wheat yields (5,200–6,900 kg ha^{-1}) with a plant density of 300 plants m^{-2} and a fertilizer N rate of 138 kg ha^{-1} for an early maturing cultivar Mekelle-3.

Effects of Water Management

Earlier simulation studies have shown that tied ridging techniques could have an equivalent impact on soil water and yield as with two irrigation applications during the mid-growth stages (Araya et al., 2021). Tied ridging practices are effective in harvesting rainwater (reducing runoff), increasing infiltration, soil water storage, and yield (Araya & Stroosnijder, 2010; Biazin & Stroosnijder, 2012). Two irrigations (simulating tide ridges) slightly improved wheat yield when compared with rainfed or dryland wheat (Figures 3.2 and 3.3). However, tied ridges might not significantly improve wheat yield in sandy clay loam soils during normal years if early planting date were used, whereas it may substantially enhance yield compared with the conventional dryland if wheat were planted late (Figures 3.2 and 3.3e–f). This study showed irrigation (tied ridging) and plant density were major factors influencing yield when planted late. Araya and Stroosnijder (2010) and Biazin and Stroosnijder (2012) indicated that the tied ridging could enhance yield during below average rainfall seasons. The relative increase in yield due to the application of tied ridging under the same planting date and plant density was less than or equal to 700 kg ha^{-1}.

Effects of N Fertilization Rates

The relationship between wheat yield and N application rate under different planting dates is presented in Figures 3.3 and 3.4. However, the increase in yield due to N rates varied with planting dates. Simulated future yield increased toward early planting with increased N application rate up to 192 kg ha^{-1} and showed slight reduction toward late planting (Figure 3.4). The positive impact of N on simulated yield was observed when the crop was planted between 20 June and 10 July of the planting window. Assuming most farmers plant wheat around 20 June with N management of zero (no N fertilizer), 16 and 64 kg N ha^{-1}, wheat yield during the mid-century increased on average by 36, 35, and 27%, respectively, compared with the corresponding baseline wheat yield of the respective N management (Figure 3.4a). The yield changes slightly decreased if most farmers would choose to plant wheat between 20 June and 10 July with N management of zero (no N fertilizer), 16 and 64 kg N ha^{-1}. In this case, wheat yield during the mid-century could increase on average by 42, 23, and 21%, respectively, compared with the corresponding baseline wheat yield of the respective N management (Figure 3.4a–c). Similarly, under the early planting strategy (20 June–10 July), increased N from 0 to 64 kg N ha^{-1} could increase wheat yield by 77 and 115% during the baseline and mid-century period, respectively (Figure 3.4a–c). The influence of N application on simulated wheat yield was reduced when the crop was planted late (1–20 August). For late planting, N rate of 100 kg ha^{-1} or above did not show significant yield improvement in dryland wheat.

In the low-input crop production systems like those that exist in Ethiopia and other Sub-Saharan Africa (SSA) regions, appropriate N management is key to improving crop productivity. Farmers apply little to no N fertilizer to wheat due to economic constraints. A recent fully comprehensive survey indicated that N deficiency was the top limiting soil fertility factor in Ethiopia (Stewart et al., 2020). Furthermore, soils in SSA are generally poor in organic C content and deficient in N and P due to continuous mining of soil nutrients through crop production and not replenishing these nutrients back through inorganic or organic sources

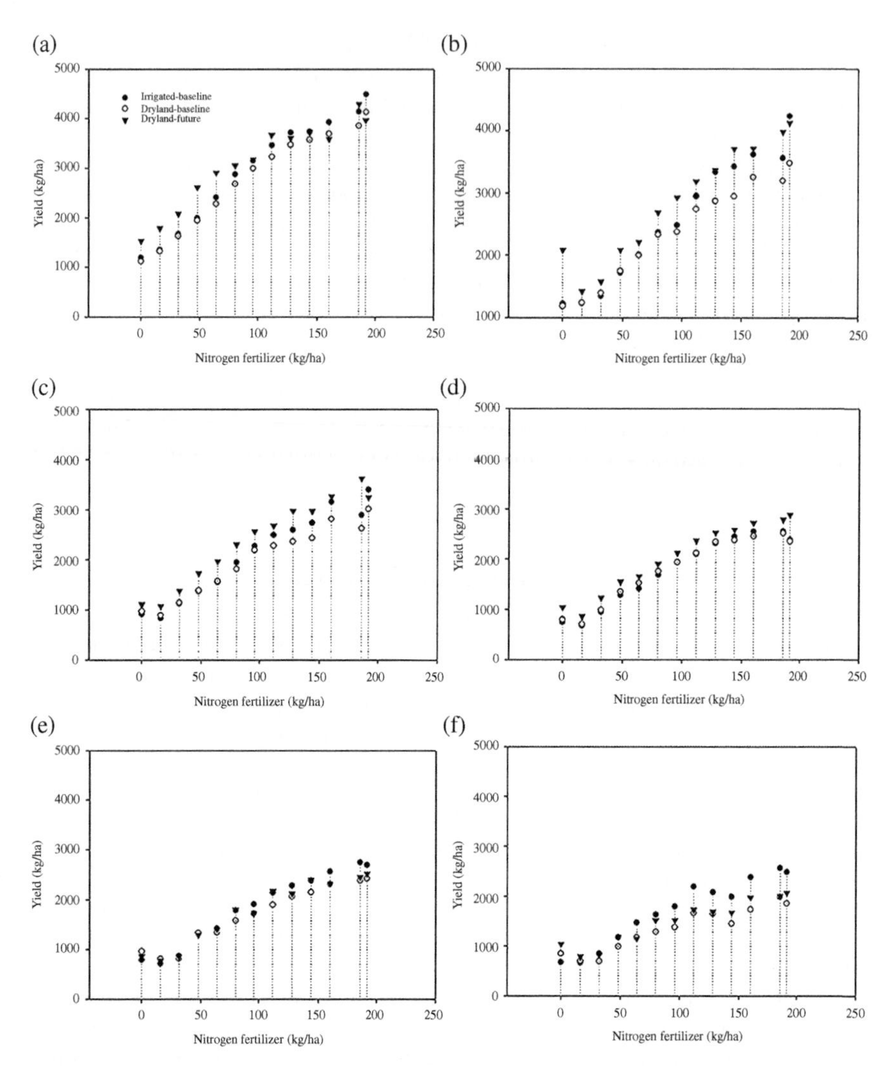

Figure 3.4 Future dryland wheat yield and irrigated and dryland baseline yields (under 300 plants m^{-2}) as affected by fertilizer N application rate under different planting dates—(a) 20 June, (b) 1 July, (c) 10 July, (d) 20 July, (e) 1 August, and (f) 10 August—in sandy clay loam soils of Adigudom, northern Ethiopia.

(Stewart et al., 2020). Overall fertilizer use in SSA is low and it needs to be improved to increase productivity of crops. This survey identified that strengthening of inorganic fertilizer-base systems and increasing access to and use of quality organic inputs are among top recommendations to improve soil fertility and crop yields (Stewart et al., 2020). Integrated nutrient management, which includes use of inorganic fertilizer along with organic sources of nutrients (such as manure, compost, rotation with legume crops, and use of agroforestry and shrub/bush and trees), retention of crop residues, including of cover crops or green manure crops, provides a holistic solution to address soil fertility issues in Ethiopia and SSA (Shepherd et al., 1995; Snapp et al., 1998; Stewart et al., 2020; Vanlauwe et al., 2014a, 2014b).

Effects of Future Climate Change under Different Crop Management Practices

For the mid-century period, simulated wheat yield under different planting date scenarios was not substantially different from baseline yield (Figures 3.5 and 3.6) in the study region. Similar to the baseline scenario, yield was substantially influenced by the selection of planting date. Therefore, as in the baseline, there was a yield penalty when planting was delayed. Assuming 0 N application as control, there was a substantial increase in yield (8–25%) for every increment (16 kg N ha^{-1}) of N (Figure 3.6). The percentage increase was modest for the future scenario due to relatively higher yield for 0 N (control) for the future scenario when compared with that simulated for the baseline scenario.

However, the average percentage increase in yield for every 16 kg N ha^{-1} increment at each planting date was simply a relative increase when compared with the control under similar planting date and planting density. Wheat yield during the baseline period increased by more than 200% at the higher N rate (>192 kg ha^{-1}) relative to the control (dryland wheat yield with 0 N) (Figure 3.7), which can significantly contribute to the overarching goal of food security in the target region. As in the baseline scenario, there was an increase in wheat yield under the future scenario by more than 200% at higher N rates (>186 kg ha^{-1})

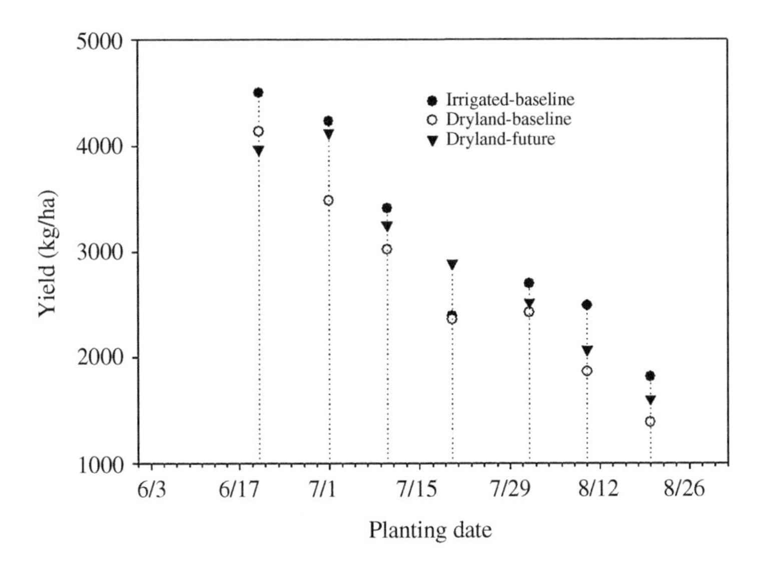

Figure 3.5 Future dryland wheat yield and irrigated and dryland baseline yield as affected by planting date under relatively high fertilizer N application rate (192 kg ha^{-1}) in sandy clay loam soils of Adigudom area, northern Ethiopia.

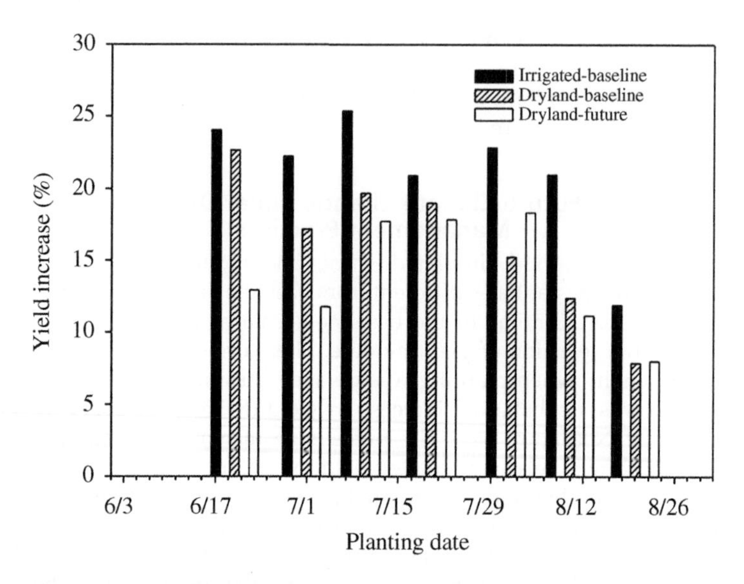

Figure 3.6 Average percentage yield increase per increase in N fertilization by 16 kg ha^{-1} for irrigated and dryland (under baseline) and for dryland future scenario when compared with the yield obtained under the corresponding 0 N fertilization for each planting date scenario, under sandy clay loam soils of Adigudom area, northern Ethiopia.

when compared with the control (dryland wheat without N fertilizer) (Figure 3.7). This study showed that increased N application rate substantially improved wheat yields (Figures 3.6 and 3.7) under future scenarios. Simulated future dryland yield was slightly higher than the corresponding dryland baseline yield (not shown). Simulated future dryland yield was also slightly higher than the irrigated baseline wheat yield (tied ridging, simulated as two irrigations, each 30 mm during flowering stages). There was a strong linear relationship between simulated future dryland ($R^2 = .91$) and irrigated yield ($R^2 = .95$) relative to baseline yield (Figure 3.8).

As mentioned for the baseline yield, much of the yield increase came from N fertilizer and early planting. The lowest simulated future yield was for low N rates and late planting. The simulated future yield was slightly higher, but it was not significantly different from the baseline yield, potentially indicating that improving agronomic management practices may adequately reduce the anticipated negative impact of climate. Nitrogen management for wheat could vary depending on fertilizer availability/supply, economic condition, and yield expectations. Amount of N applied by wheat growers range from 0 kg N ha^{-1} (no fertilizer) to 32 kg N ha^{-1} with an average of 16 kg N ha^{-1} (Araya et al., 2017a). However, the recommended N fertilizer rate for the study area is approximately 64 kg N ha^{-1} (100 kg diammonium phosphate [DAP]; [46% P and 18% N] and 100 kg Urea ha^{-1} [46% N ha^{-1}]) (Araya et al., 2017a). Accordingly, Araya et al. (2017a) simulated yield based on farmers' prevailing N management (0–32 kg N ha^{-1}) for approximately 200 farmers during the baseline period (1980–2009) was compared with wheat grown under recommended N management (64 kg N ha^{-1}) during the

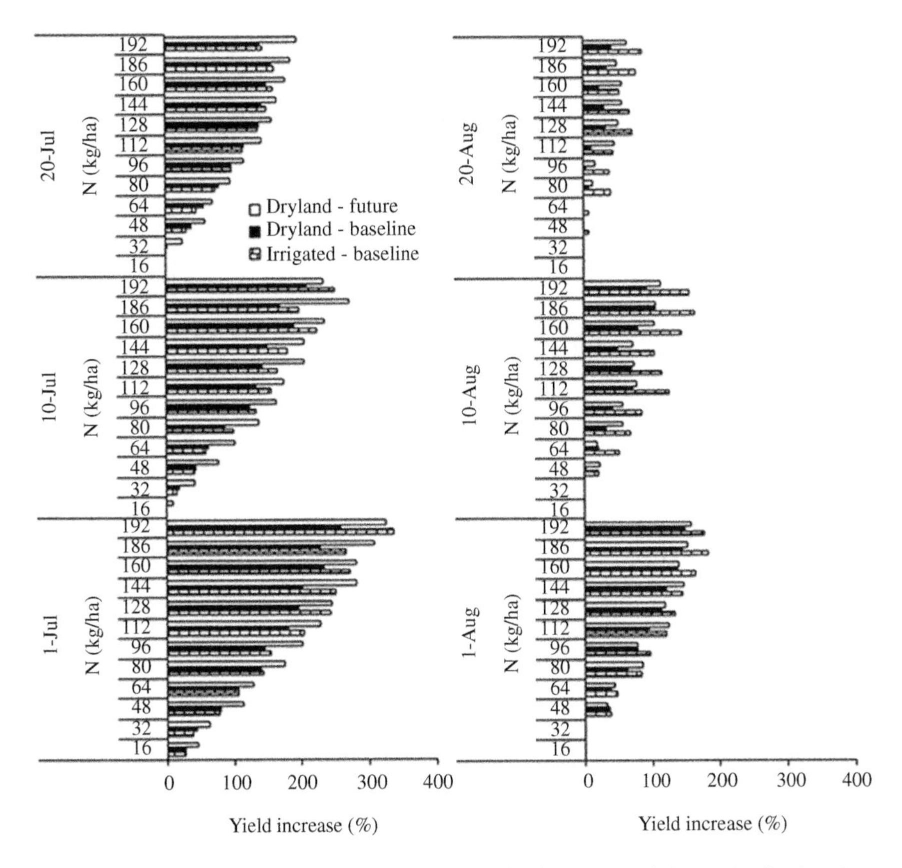

Figure 3.7 Relative yield increase due to increased N fertilizer rate and change in planting date when compared with control (average yield for 0 N kg ha^{-1} under different planting dates) for baseline and future dryland and irrigated scenarios, under sandy clay loam soils of Adigudom area, northern Ethiopia.

mid-century period. They found that future yield based on recommended fertilizer (64 kg N ha^{-1}) increased by more than 100% compared with the baseline yield based on baseline management (16 kg N ha^{-1}) (Araya et al., 2017a). The findings partially agreed well with the current work, although the magnitude of increase substantially varied by planting date. Significant amount of the yield increase was simulated from increased N fertilizer. Evaluating rate of increase in yield per N fertilizer rate was an important technique to determine yield based on a given fertilizer rate per given planting date and planting density under the baseline and future climate. This modeling technique can give valuable information for implementing precision agriculture practices and for developing predictive farming tools. This information can help reduce operational costs and enhance input use efficiency. However, the research was based on climate data input from two Global Climate Models (GCMs) and one crop model. Asseng et al. (2015) used 30 different models and reported that the utilization of multiple model ensemble

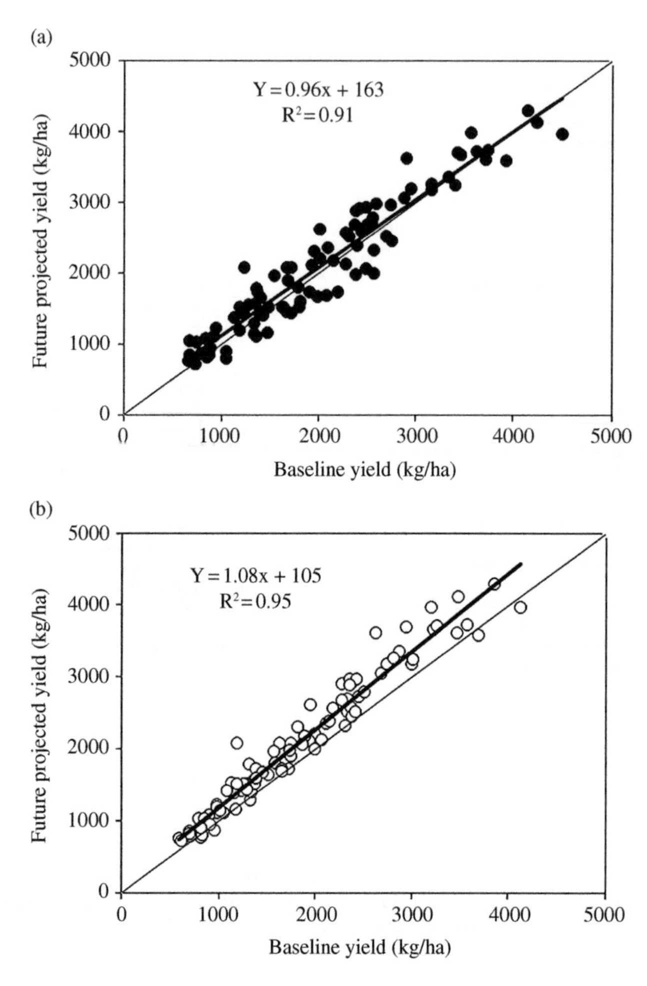

Figure 3.8 Relationship between future dryland wheat yield and simulated (a) irrigated and (b) dryland baseline yields under various agronomic management in sandy clay loam soils of Adigudom, northern Ethiopia.

might improve the accuracy of yield prediction under high temperatures. Similarly, Liu et al. (2016) showed that grid-based and point-based simulations and statistical regressions (from historic data) projected similar impacts of high temperature stress; however, they conclude that by using multi-method ensemble can help quantify the method uncertainty.

Overall, this study showed that the increase in temperature did not go above the temperature that cause significant yield reductions. Wheat responded to a change in climate similar to that of the baseline (20.3 °C), implying that the increase in temperature in the mid-century stayed within the optimum range for the wheat cultivar, negative effect of increased temperature was compensated by the beneficial effect of increased CO_2, and the change in rainfall was not significant. Reports showed that temperatures above the optimum requirement of the crop could reduce yield (Hatfield et al., 2011; Prasad et al., 2017). Optimum temperature for anthesis and grain filling of wheat are 21 and 20.7 °C, respectively (Porter & Gawith, 1999), whereas others reported optimum temperature for grain development could range between 15 and 25 °C (Nuttall et al., 2017), which is a relatively wider range. In addition, the expected rise in future CO_2 level (571 µmol mol^{-1}) and seasonal rainfall for mid-century under RCP8.5 was anticipated to benefit the wheat crop and stabilize yield under the future scenario where temperatures are below the maximum and under optimal range for crop growth, development, and yield formation.

Finally, it is not only important to identify the best management options to overcome negative impacts of climate change under low-input agriculture, but also to understand the barriers of adoption by resource-poor farmers and create enabling environments and policies for adoption of innovative practices. Addressing the complex issues of climate change requires participatory approaches involving multidisciplinary teams to identify research priorities, using innovative approaches, and developing holistic solutions (Middendorf et al., 2020).

Conclusions

Early planting with a planting density of 300 plants m^{-2} maximized wheat yields in the target region of northern Ethiopia. Wheat responded positively to increasing N application. There was an increase in yield ranging from 8 to 25% for every increase of 16 kg N ha^{-1} until about 180 kg N ha^{-1}. Considering most farmers would plant wheat within a month period (20 June–20 July) under 0, 16, and 64 kg N ha^{-1} management scenarios, mid-century wheat yield increased on average by 36, 23, and 18%, respectively, compared with the corresponding baseline wheat yield of the respective N management. Similarly, increased N from 0 to 64 kg N ha^{-1} could increase wheat yield by 81 and 114% under the baseline and mid-century periods, respectively.

Irrigation treatments (two irrigations equivalent to that saved by tied ridging at reproductive stages of crop development) resulted in higher yields under all management practices compared with dryland. There were no negative impacts of future climate scenarios for mid-century (2040–2069) on wheat yield under different management practices. This suggested that the increase in temperature in the mid-century stayed within the optimum range for the wheat cultivar, negative effect of increased temperature was compensated by the beneficial effect of increased CO_2, and the change in rainfall was not significant. However, we acknowledge that in our simulations we did not consider the occurrence of extreme environmental events (short episodes of high temperature and drought stresses) on yield, which are likely to increase in future climates. Similarly, the impact of various biotic stresses (pests, disease, and weeds) were not considered. In this exercise, we also did not quantify the economic benefits of different management practices (particularly N), which is important for smallholder farmers for adoption and scaling. Some crop models including DSSAT can conduct analyses of cost–benefit ratio and economic returns, based on market prices, and these should be considered in future simulations or research. Overall, this exercise provides an example for how crop models can be used for identifying suitable management practices and to forecast potential impact of climate change on crop yields. Crop models can provide initial analyses of available options that can be tested and validated under field conditions before adoption and scaling. Such information will be useful to agronomists, plant biologists, and policy makers for development and scaling of suitable technologies for increasing and/or sustaining yield under current and future changing environments and socio-economic and management conditions.

Acknowledgments

The authors would like to thank the Feed the Future Innovation Lab for Collaborative Research on Sustainable Intensification (Grant no. AID-OAA-L-14-00006) funded by the US Agency for International Development. We would like to thank Atkilt Girma for his support in generating the future climate scenarios. Contribution no. 19-336-B from the Kansas Agricultural Experiment Station.

Abbreviations

APSIM, Agricultural Production System sIMulator; DAP, diammonium phosphate; DSSAT–CSM, Decision Support System for Agrotechnology–Cropping Systems Model; GCMs, Global Climate Models; RCP, Representative Concentration Pathways; SSA, Sub-Saharan Africa.

References

AgMIP. (2013). *Guide for running AgMIP climate scenario generation tools with R in windows, version 2.3.* https://docplayer.net/56928429-Guide-for-running-agmip-climate-scenario-generation-tools-with-r.html

Araya, A., Hoogenboom, G., Luedeling, E., Hadgu, K. M., Kisekka, I., & Martorano, L. G. (2015). Assessment of maize growth and yield using crop models under present and future climate in southwestern Ethiopia. *Agricultural and Forest Meteorology, 214/215,* 252–265.

Araya, A., Keesstra, S. D., & Stroosnijder, L. (2010). A new agro-climatic classification for crop suitability zoning in northern semi-arid Ethiopia. *Agricultural and Forest Meteorology, 150,* 1047–1064.

Araya, A., Kisekka, I., Girma, A., Hadgu, K. M., Beltrao, N. E. S., Ferreira, H., Afewerk, A., Birhane, A., Tsehaye, Y., & Martorano, L. G. (2017a). The challenges and opportunities for wheat production under future climate in northern Ethiopia. *The Journal of Agricultural Science, 155,* 379–393. https://doi.org/10.1017/S0021859616000460

Araya, A., Kisekka, I., Gowda, P. H., & Prasad, P. V. V. (2017b). Evaluation of water-limited cropping systems in a semi-arid climate using DSSAT–CSM. *Agricultural Systems, 150,* 86–98.

Araya, A., Kisekka, I., Prasad, P. V. V., Holman, J., Foster, A. J., & Lollato, R. (2017c). Assessing wheat yield, biomass, and water productivity responses to growth stage based irrigation water allocation. *Transactions of the American Society of Agricultural and Biological Engineers, 60,* 107–121. https://doi.org/10.13031/trans.11883

Araya, A., Prasad, P. V. V., Ciampitti, I. A., & Jha, P. K. (2021). Using crop simulation model to evaluate influence of water management practices and multiple cropping systems on crop yields: A case study for Ethiopian highlands. *Field Crops Research, 260,* 108004. https://doi.org/10.1016/j.fcr.2020.108004

Araya, A., Prasad, P. V. V., Gowda, P. H., Afewerk, A., Abadid, B., & Foster, A. J. (2019a). Modeling irrigation and nitrogen management of wheat in northern Ethiopia. *Agricultural Water Management, 216,* 264–272. https://doi.org/10.1016/j.agwat.2019.01.014

Araya, A., Prasad, P. V. V., Gowda, P. H., Kisekka, I., & Foster, A. J. (2019b). Yield and water productivity of winter wheat under various irrigation capacities. *Journal of the American Water Resources Association, 55,* 24–37. https://doi.org/10.1111/1752-1688.12721

Araya, A., Prasad, P. V. V., Gowda, P. H., Djanaguiraman, M., & Kassa, A. H. (2020a). Potential impact of climate change factors and agronomic adaptation strategies on wheat yield in central highlands of Ethiopia. *Climate Change, 159,* 461–479.

Araya, A., Prasad, P. V. V., Zambreski, Z., Gowda, P. H., Ciampitti, I. A., Assefa, Y., & Girma, A. (2020b). Spatial analysis of the impact of climate change factors and adaptation strategies on productivity of wheat in Ethiopia. *The Science of the Total Environment, 731*, 139094. https://doi.org/10.1016/j.scitotenv.2020.139094

Araya, A., Gowda, P. H., Golden, B., Foster, A. J., Aguilar, J., Currie, R., Ciampitti, I. A., & Prasad, P. V. V. (2019c). Economic value and water productivity of major irrigated crops in the Ogallala aquifer region. *Agricultural Water Management, 214*, 55–63. https://doi.org/10.1016/j.agwat.2018.11.015

Araya, A., & Stroosnijder, L. (2010). Effects of tied ridges and mulch on barley (*Hordeum vulgare*) rainwater use efficiency and production in Northern Ethiopia. *Agricultural Water Management, 97*, 841–847. https://doi.org/10.1016/j.agwat.2010.01.012

Araya, A., & Stroosnijder, L. (2011). Assessing drought risk and irrigation need in northern Ethiopia. *Agricultural and Forest Meteorology, 151*, 425–436.

Araya, A., Stroosnijder, L., Solomon, H., Keesstra, S. D., Berhe, M., & Hadgu, K. M. (2012). Risk assessment by sowing date for barley (*Hordeum vulgare*) in northern Ethiopia. *Agricultural and Forest Meteorology, 154/155*, 30–37.

Asseng, S., Foster, I., & Turner, N. C. (2011). The impact of temperature variability on wheat yields. *Global Change Biology, 17*, 997–1012. https://doi.org/10.1111/j.1365-2486.2010.02262.x

Asseng, S., Ewert, F., Martre, P., Rötter, R. P., Lobell, D. B., Cammarano, D., Kimball, B. A., Ottman, M. J., Wall, G. W., White, J. W., Reynolds, M. P., Alderman, P. D., Prasad, P. V. V., Aggarwal, P. K., Anothai, J., Basso, B., Biernath, C., Challinor, A. J., De Sanctis, G., . . . Zhu, Y. (2015). Rising temperatures reduce global wheat production. *Nature Climate Change, 5*, 143–147. https://doi.org/10.1038/nclimate2470

Barlow, K. M., Christy, B. P., O'Leary, G. J., Riffkin, P. A., & Nuttall, J. G. (2015). Simulating the impact of extreme heat and frost events on wheat crop production: A review. *Field Crops Research, 171*, 109–119. https://doi.org/10.1016/j.fcr.2014.11.010

Biazin, B., & Stroosnijder, L. (2012). To tie or not to tie ridges for water conservation in Rift Valley drylands of Ethiopia. *Soil Tillage Research, 124*, 83–94.

Boote, K. J., Allen, L. H., Prasad, P. V. V., Baker, J. T., Gesch, R. W., Snyder, A. M., Pan, D., & Thomas, J. M. G. (2005). Elevated temperature and CO_2 impacts on pollination, reproductive growth, and yield of several globally important crops. *Journal of Agricultural Meteorology, 60*, 469–474. https://www.doi.org/10.2480/agrmet.469

Boote, K. J., Allen, L. H., Prasad, P. V. V., & Jones, J. W. (2011). Testing effects of climate change in crop models. In D. Hillel & C. Rosenzweig (Eds.), *Handbook of climate change and agroecosystems* (pp. 109–129). Imperial College Press.

Boote, K. J., Jones, J. W., Hoogenboom, G., & White, J. W. (2012). *The role of crop systems simulation in agriculture and environment*. IGI Global. https://doi.org/10.4018/978-1-4666-0333-2.ch018

Boote, K. J., Prasad, P. V. V., Allen, L. H., Singh, P., & Jones, J. W. (2018). Modeling sensitivity of grain yield to elevated temperature in the DSSAT crop models for peanut, soybean, drybean, chickpea, sorghum, and millet. *European Journal of Agronomy, 100*, 99–109.

Craufurd, P. Q., Vadez, V., Jagadish, S. V. K., Prasad, P. V. V., & Zaman-Allah, M. (2013). Crop science experiments designed to inform crop modeling. *Agricultural and Forest Meteorology, 170*, 8–18.

CSA. (2000–2018). Agricultural sample survey 2000/2001–2018. Report on Area and Production of Crops (Private Peasant Holdings, Meher Season). In *Statistical Bulletin*. Central Statistical Agency.

Dunne, J. P., John, J. G., Adcroft, A. J., Griffies, S. M., Hallberg, R. W., Shevliakova, E., Stouffer, R. J., Cooke, W., Dunne, K. A., Harrison, M. J., Krasting, J. P., Malyshev, S. L., Milly, P. C. D., Phillipps, P. J., Sentman, L. T., Samuels, B. L., Spelman, M. J., Winton, M., Wittenberg, A. T., & Zadeh, N. (2012). GFDL's ESM2 global coupled climate–carbon earth system models. Part I: Physical formulation and baseline simulation characteristics. *Journal of Climate, 25*, 6646–6665. https://doi.org/10.1175/JCLI-D-11-00560.1

Hatfield, J. L., Boote, K. J., Kimball, B. A., Ziska, L. H., Izaurralde, R. C., Ort, D., Thomson, A. M., & Wolfe, D. (2011). Climate impacts on agriculture: Implications for crop production. *Agronomy Journal, 103*, 351–370. https://doi.org/10.2134/agronj2010.0303

Hoogenboom, G., Jones, J. W., Wilkens, P. W., Porter, C. H., Boote, K. J., Hunt, L. A., Singh, U., Lizaso, J. I., White, J. W., Uryasev, O., Ogoshi, R., Koo, J., Shelia, V., & Tsuji, G. Y. (2015). *Decision Support System for Agrotechnology Transfer (DSSAT) Version 4.6*. DSSAT Foundation.

IPCC. (2014). Climate change 2014: Synthesis report. In R. K. Pachauri & L. A. Meyer (Eds.), *Contribution of working groups I, II and III to the fifth assessment report of the intergovernmental panel on climate change*. IPCC.

Jones, C. D., Hughes, J. K., Bellouin, N., Hardiman, S. C., Jones, G. S., Knight, J., Liddicoat, S., O'Connor, F. M., Andres, R. J., Bell, C., Boo, K.-O., Bozzo, A., Butchart, N., Cadule, P., Corbin, K. D., Doutriaux-Boucher, M., Friedlingstein, P., Gornall, J., Gray, L., . . . Zerroukat, M. (2011). The HadGEM2-ES implementation of CMIP5 centennial simulations. *Geoscientific Model Development, 4*, 543–570. https://doi.org/10.5194/gmd-4-543-2011

Jones, J. W., Hoogenboom, G., Porter, C. H., Boote, K. J., Batchelor, W. D., Hunt, L. A., & Ritchie, J. T. (2003). The DSSAT cropping system model. *European Journal of Agronomy, 18*, 235–265.

Jones, P. G., & Thornton, P. K. (2003). The potential impacts of climate change on maize production in Africa and Latin America in 2055. *Global Environmental Change, 13*, 51–59.

Joshi, N., Singh, A. K., & Madramootoo, C. A. (2015). Corn yield simulation under different nitrogen loading and climate change scenarios. *Journal of Irrigation and Drainage Engineering, 141*(10), 04015013.

Liu, B., Asseng, A., Muller, C., Ewert, F., Elliott, J., Lobell, D. B., Martre, P., Ruane, A. C., Wallach, D., Jones, J. W., Rosenzweig, C., Aggarwal, P. K., Alderman, P. D., Anothai, J., Basso, B., Biernath, C., Cammarano, D., Challinor, A., Deryng, D., . . . Zhu, Y. (2016). Similar estimates of temperature impacts on global wheat yield by three independent methods. *Nature Climate Change, 6*, 1130–1136. https://doi.org/10.1038/nclimate3115

Majaivana, D. D., Magwende, L. I., & Olonga, H. F. (2016). Determinant of optimum plant density, nitrogen and phosphorus fertilization for different wheat genotypes. *International Journal of Manures and Fertilizers, 4*, 592–599.

Middendorf, B. J., Prasad, P. V. V., & Pierzynski, G. M. (2020). Setting research priorities for tackling climate change. *Journal of Experimental Botany, 71*, 480–489.

Nuttall, J. G., O'Leary, G. J., Panozzo, J. F., Walker, C. K., Barlow, K. M., & Fitzgerald, G. J. (2017). Models of grain quality in wheat. A review. *Field Crops Research, 202*, 136–145. https://doi.org/10.1016/j.fcr.2015.12.011

Porter, J. R., & Gawith, M. (1999). Temperatures and the growth and development of wheat: A review. *European Journal of Agronomy, 10*, 23–36.

Prasad, P. V. V., Bhemanahalli, R., & Jagadish, S. V. K. (2017). Field crops and the fear of heat stress: Opportunities, challenges and future directions. *Field Crops Research, 200*, 114–121. https://doi.org/10.1016/j.fcr.2016.09.024

Prasad, P. V. V., Boote, K. J., & Allen, L. H., Jr. (2006). Adverse high temperature effects on pollen viability, seed-set, seed yield and harvest index of grain-sorghum [*Sorghum bicolor* (L.). Moench] are more severe at elevated carbon dioxide due to higher tissue temperatures. *Agricultural and Forest Meteorology, 139*, 237–251.

Prasad, P. V. V., & Djanaguiraman, M. (2014). Response of floret fertility and individual grain weight of wheat to high temperature stress: Sensitive stages and thresholds for temperature and duration. *Functional Plant Biology, 41*, 1261–1269. https://doi.org/10.1071/FP14061

Prasad, P. V. V., Djanaguiraman, M., Perumal, R., & Ciampitti, I. A. (2015). Impact of high temperature stress on floret fertility and individual grain weight of grain sorghum: Sensitive stages and thresholds for temperature and duration. *Frontiers in Plant Science, 6*, 820. https://doi.org/10.3389/fpls.2015.00820

Rosenzweig, C., Antle, J. M., Ruane, A. C., Jones, J. W., Hatfield, J., Boote, K. J., Thorburn, P., Valdivia, R. O., Descheemaeker, K., Porter, C. H., Janssen, S., Bartels, W. L., Sulivan, A., & Mutter, C. Z. (2016). *Protocols for AgMIP Regional Integrated Assessments Version 7.0*. https://agmip.org/wp-content/uploads/2018/08/AgMIP-Protocols-for-Regional-Integrated-Assessment-v7-0-20180218-1-ilovepdf-compressed.pdf

Ruane, A. C., Cecil, L. D., Horton, R. M., Gordónd, R., McCollume, R., Browne, D., Killough, B., Goldberg, R., Greeley, A. P., & Rosenzweig, C. (2013). Climate change impact uncertainties for maize in Panama: Farm information, climate projections, and yield sensitivities. *Agricultural and Forest Meteorology, 170*, 132–145.

Shepherd, K. D., Ohlsson, E., Okalebo, J. R., & Ndufa, J. K. (1995). Potential impact of agroforestry on soil nutrient balance at the farm scale in the East Africa highlands. *Fertilizer Research, 44*, 87–99.

Silungwe, F. R., Graef, F., Bellingrath–Kimura, S. D., Tumbo, S. D., Kahimba, F. C., & Lana, M. A. (2018). Crop upgrading strategies and modelling for rainfed cereals in a semi-arid climate: A review. *Water, 10*, 356.

Singh, P., Nedumaran, S., Traore, P. S., Boote, K. J., Rattunde, H. F. W., Prasad, P. V. V., Singh, N. P., Srinivas, S., & Bantilan, M. C. S. (2015). Quantifying potential benefits of drought and heat tolerance in rainy season sorghum for adapting to climate change. *Agricultural and Forest Meteorology, 185*, 37–48.

Snapp, S. S., Mafongoya, P. L., & Waddington, S. (1998). Organic matter technologies for integrated nutrient management in smallholder cropping systems of southern Africa. *Agriculture, Ecosystems & Environment, 71*, 185–200. https://doi.org/10.1016/S0167-8809(98)00140-6

Stewart, Z. P., Pierzynski, G. M., Middendorf, B. J., & Prasad, P. V. V. (2020). Approaches to improve soil fertility in sub-Saharan Africa. *Journal of Experimental Botany, 71*, 632–641.

Stroosnijder, L. (2009). Modifying land management in order to improve efficiency of rainwater use in the African highlands. *Soil Tillage Research, 103*, 247–256.

Taffesse, A. S., Dorosh, P., & Asrat, S. (2012). *Crop production in Ethiopia: Regional patterns and trends*. ESSP II Working Paper no. 016. Development Strategy and Governance Division, International Food Policy Research Institute, Ethiopia Strategy Support Program II.

Vanlauwe, B., Descheemaeker, K., Giller, K. E., Huising, J., Merckz, R., Nziguiheba, G., Wendt, J., & Zingore, S. (2014a). Integrated soil fertility management in sub-Saharan Africa: Unravelling local adaption. *Soil, 1*, 491–508.

Vanlauwe, B., Wendt, J., Giller, K. E., Corbeels, M., Gerard, B., & Nolte, C. (2014b). A fourth principle is required to define conservation agriculture in sub-Saharan Africa: The appropriate use of fertilizer to enhance crop productivity. *Field Crops Research, 155*, 10–13. http://dx.doi.org/10.1016/j.fcr.2013.10.002

4

Dilys S. MacCarthy,
Samuel G. K. Adiku,
Alpha Y. Kamara,
Bright S. Freduah, and Joseph X. Kugbe

The Role of Crop Simulation Modeling in Managing Fertilizer Use in Maize Production Systems in Northern Ghana

Abstract

The agricultural landscape in Sub-Sahara Africa is dominated by smallholders and characterised by high heterogeneity due to high variation in soil, crop varieties and land management practices. As a consequence, crop yields vary considerably in space and time. This study aimed at developing a production domain for different maize varieties in northern Ghana in response to the conditions, using a simulation modelling approach. Data from a two-year multi-location nitrogen (N) response trial in northern Ghana were used to calibrate and evaluate the performance of the Decision Support System for Agro-technological Transfer (DSSAT) crop model, which was then used to simulate maize yields across the region. The model performed adequately in simulating maize phenology, with the relative root mean square error (RRMSE) below 15%. The response of grain and biomass yield of different varieties of maize to nitrogen fertilizer application was also adequately simulated, with model efficiency coefficients ranging between .68 and .88 across the varieties and planting dates. In general, grain yields were lower in the Sudan Savannah than in the Guinea Savanna zones of northern Ghana. Also, mid planting produced higher yields in the Sudan Savanna zone whereas for the Guinean zone, it was the early and late planting that produced higher maize yields. The simulations showed that the application of 60 kg N ha^{-1} was optimal for the early and the extra-early varieties, but higher N application rate of 90 kg N ha^{-1} was optimal for the intermediate variety. It is concluded that the DSSAT crop model enabled an effective generation of crop yield data suitable for decision support, agricultural planning and policy formulation in northern Ghana.

Enhancing Agricultural Research and Precision Management for Subsistence Farming by Integrating System Models with Experiments, First Edition. Edited by Dennis J. Timlin and Saseendran S. Anapalli.

DOI: 10.1002/9780891183891.ch04

Introduction

Agriculture in Ghana and in general, Sub-Saharan Africa is dominated mainly by subsistence and rain-fed systems (MacCarthy et al., 2018). The productivity of the soils is low and food production is bedeviled with several constraints. Prominent among the constraints are poor soil fertility and erratic rainfall distribution. The fertility of the soils has been on a decline over the decades due to continuous cultivation with minimal nutrient application. In addition, the fertility of the soils continues to deteriorate because the traditional bush fallow systems (shifting cultivation), which serve to restore soil nutrients and soil organic matter, have become ineffective due to the shortening of the fallow periods. The distribution of rainfall is also increasingly becoming more erratic over the growing season, making investments into agriculture a risky venture. Thus, most financial institutions are reluctant to provide credit to farmers for the purposes of investing in their farming activities. Another major cause of low yield is due to the low quality of planting materials that are in use. Most of the crop varieties are low yielding and susceptible to other abiotic stress factors. Although improved and high-yielding varieties have been released by institutions such as the International Institute for Tropical Agriculture, Ibadan, Nigeria, and are available to farmers, yields continue to be constrained by low-level soil fertility and low fertilizer application.

In the Guinea Savanna agro-ecology where some of the new maize (*Zea mays* L.) varieties have been introduced, there is a wide spatial variation in the soils and weather characteristics. This region constitutes about 41% of Ghana's total land area (97,702 km^2) and is deemed to become a major basket for cereal production in the country. The yield of any maize variety will depend on the nature of soil and topographical characteristics, rainfall distribution, and other crop management practices. To be able to adequately provide sound recommendations to farmers on varietal choice, or management practices that are suitable for their specific location, long-term field observations that have to be replicated all over the entire region will be required. This can be time consuming and very expensive. Cropping systems simulation models provide the opportunity to achieve these based on limited experimentation. The Decision Support Systems for Agro-Technological Transfer (DSSAT) crop simulation model, which has been validated in this region, can be used to assess varietal responses to abiotic stresses. The DSSAT model simulates the yield and growth of crops based on daily weather (minimum and maximum temperature, rainfall, and solar radiation), soil profile information, site information, and crop management practices. Although the model has been variously used to inform and guide decision-making in maize production systems globally, its use across northern Ghana is currently limited. This hinders efforts made to increase maize yield and productivity across the region, and for predictive purposes in the inherently low nutrient soils and erratic rain-fed systems. A good predictive model can also reinvigorate financial investments in the mostly resource-poor, subsistence production system. The overall aim in this study is to develop production domains for different maize varieties in northern Ghana. To achieve this, the following specific objectives were set out: (a) derive the appropriate genetic coefficients for new and existing maize varieties for calibrating the DSSAT–Crop Environment

REsource Synthesis (CERES) maize model; (b) evaluate the response of the maize model to mineral N fertilizer application; (c) determine appropriate fertilizer recommendation for maize varieties in northern Ghana; and (d) derive maize yield maps for northern Ghana based on management factors such as planting date, fertilizer application rates, maize varieties, among other factors that can guide farm decision making.

Methodology

Study Area

The northern regions of Ghana lie in the Guinea and Sudan Savanna agro-ecologies (Kugbe et al., 2015). The vegetation is mainly grassland with scattered trees. The climate is characterized by an annual average temperature of 32 °C. Average rainfall is 1,100 mm, distributed in a mono-modal rainfall pattern beginning in May and spanning between 5 and 6 mo until October, followed by 5–6 mo of dry conditions. The vegetation is wooded savanna with dominant grass vegetation. Although pristine sites are fairly fertile, these are lost rapidly once converted to agriculture. The soils are coarse-textured and cultivated sites are shallow (<50 cm) and generally poor in soil organic C (<0.5%), with Fe and Al concretions present at depths of about 30 cm.

Field Experiments

In this study, several field studies were conducted at Tamale and Yendi environs in the Northern Region. Additionally, one trial was set up at the Soil and Irrigation Research Centre of the University of Ghana, Kpong (southern Ghana), for the purposes of calibration of the DSSAT model.

Even though one maize variety, Obatanpa, was previously calibrated and validated for the northern regions of Ghana (Dzotsi et al., 2010; Fosu et al., 2012), the current study involved other varieties, namely, Omankwa and Abontem. These three varieties represent typical extra early maturity group (Abontem), early group (Omankwa), and intermediate maturing group (Obatanpa). Some of the trials at the sites were carried out to provide data for calibrating these other varieties. The calibration trials involved maize growth under optimal water and N conditions at Golinga (in the years 2015 and 2016), Nyankpala (in 2016), and Kpong (in 2016). The trial involved the application of 5 Mg ha^{-1} compost 2 wk prior to planting and mineral fertilizer application at a rate of 120–45–45 kg ha^{-1} using N–P–K (15–15–15) and sulfate of ammonia at Kpong. The plant spacing used was 75 by 40 cm on plot sizes of 6 by 6 m (36 m^2). The varieties served as the treatments and were randomized in four blocks. Planting dates varied for the northern locations with the earliest in the last week of May. Data collected included crop phenology (date of emergence, date of anthesis, and date of maturity), grain yield, and biomass yield. Soil profile data (depths 0–15, 15–30, 30–45, 45–60, 60–75, and 75–100 cm) were also collected in each of the experiments. Soil samples were analyzed for particle-size distribution, organic C, and pH. Undisturbed samples were also taken to determine bulk density for each layer. A pedo-transfer function by Saxton and Rawls (2006) embedded in the DSSAT shell was used to estimate the water characteristics of the soil (field capacity, wilting point, and saturated water content). Weather data for the experimental period were obtained from weather stations close to the experimental sites.

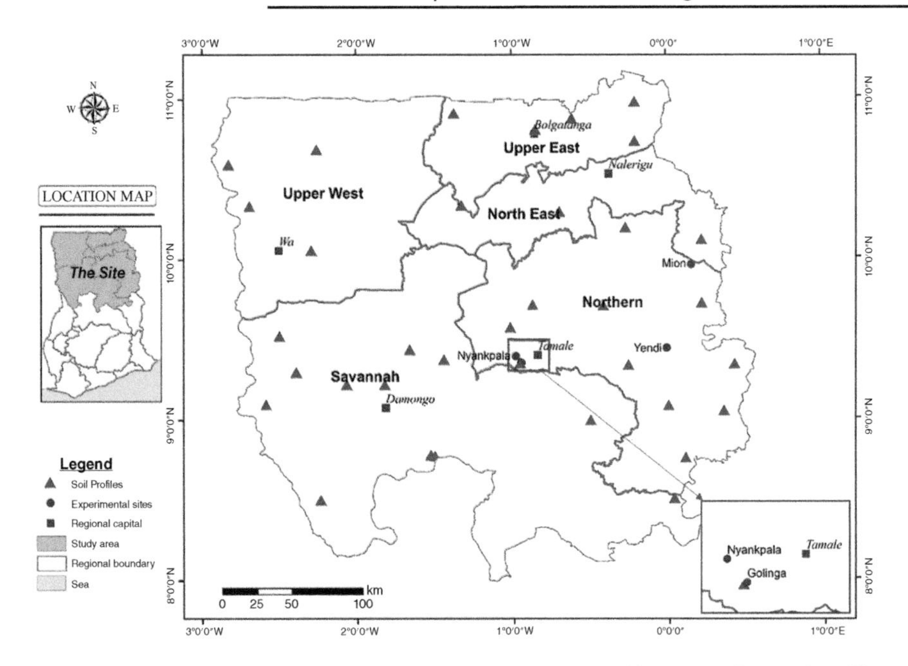

Figure 4.1 Map of northern Ghana indicating experimental sites and locations where soil profile data were collected.

Apart from trials for model calibration, another set of trials were conducted to provide independent data sets for model evaluation. These trials were conducted at Nyankpala, Golinga, and Yendi, all in the Northern Region of Ghana (Figure 4.1). The three maize varieties mentioned were planted in a split plot design with the fertilizer level being the main plot and the varieties being the sub-plots. Each of the experiments was replicated four times. The N levels used were 0, 60, 90, and 120 kg ha^{-1} and were applied as splits in the ratio 2:3 at 10 and 30 d after emergence. Phosphorous and K were also applied at the rate of 45 kg ha^{-1} P_2O_5 and K_2O at 10 d after emergence. Weed control was achieved by manual means. Data collected include date of emergence, date of 50% flag leaf stage, date of 50% flowering, date to 50% physiological maturity, and relative leaf appearance rate. Final grain yield and total biomass were taken from an area of 3 by 3 m, oven-dried to constant temperature at 70 °C and extrapolated to a kilogram-per-hectare basis. Soil samples from soil profile (0–15, 15–30, 30–45, 45–60, 60–75, and 75–100 cm) were taken; air-dried; and analyzed for pH, organic C, and particle-size distribution. Undisturbed samples were also taken from each profile to determine bulk density (Table 4.1). Weather data (daily rainfall, temperature [minimum and maximum], and solar radiation) for the various sites were collected from the Meteorological Agency of Ghana and also from the Savannah Agricultural Research Institute (SARI) in Nyankpala, Ghana.

Data collected on soils, climate, and crop management were converted into standard DSSAT file formats for genetic coefficients estimation of maize varieties and for evaluating the performance of the model to reproduce measured data collected from independent trials.

Table 4.1 Soil data used in the model

Location	L	LL	DUL	SAT	BD	Silt	Clay	OC	Soil pH
	cm	–cm³ cm⁻³–			g cm⁻³	–%–			
Golinga	15	0.085	0.176	0.35	1.49	28	32.2	0.8	6.4
	30	0.085	0.176	0.35	1.5	23	29	0.6	6.2
	45	0.127	0.192	0.36	1.5	29	18	0.5	5.8
	60	0.125	0.192	0.36	1.53	19	16	0.2	5.5
	75	0.131	0.232	0.361	1.53	28	12	0.2	5.3
	100	0.131	0.232	0.361	1.53	22	12	0.2	5.5
Nyankpala	5	0.095	0.257	0.359	1.39	18.2	7	0.41	6.7
	15	0.06	0.227	0.359	1.39	18.2	7	0.41	6.7
	30	0.06	0.228	0.34	1.59	24.5	11.3	0.32	6.4
	45	0.105	0.229	0.342	1.59	27.9	15	0.28	6.1
	60	0.12	0.205	0.342	1.63	27.5	15.5	0.28	6.1
	90	0.13	0.2	0.347	1.5	27.7	18.2	0.17	6
Yendi	5	0.05	0.175	0.359	1.47	30	8.5	0.75	5.5
	15	0.052	0.182	0.359	1.59	35	8.5	0.65	5.7
	30	0.052	0.188	0.359	1.61	40	9	0.56	6.2
	45	0.153	0.192	0.38	1.63	40	10	0.44	6.5
	60	0.167	0.198	0.4	1.63	40	10	0.2	6.5
Mion	15	0.081	0.173	0.359	1.49	32	28	0.7	6.6
	30	0.083	0.174	0.359	1.51	30	22	0.5	6.2
	45	0.122	0.19	0.36	1.53	18	29	0.4	5.8
	60	0.126	0.19	0.36	1.55	4	19	0.3	5.5
	75	0.129	0.231	0.361	1.55	12	28	0.3	5.3
	100	0.13	0.232	0.361	1.56	8	22	0.3	5.5

Note. L, depth; LL, lower limit; DUL, drained upper limit; SAT, saturated moisture content; BD, bulk density; OC, organic carbon.

Model Description

The DSSAT maize model version 4.6.0.18 (Jones et al., 2003) was used in this study. In the model, optimum yield is simulated as a function of solar radiation when other conditions such as nutrients and moisture are not limiting. The model utilizes soil profile data, crop genetic information, daily weather data, and crop management information to simulate the growth and yield of maize. The soil water balance module describes soil moisture as a function of precipitation, irrigation, soil evaporation, plant water uptake, surface runoff, and drainage. Ritchie's cascading water balance method was used to describe the movement of water between layers (Ritchie, 1998). Availability of N is through fertilizer application, the mineralization of soil organic matter, and manure applied, which are managed by the N module and the CENTURY soil organic matter module.

To use the DSSAT model for yield prediction for each maize variety required that their physiological and genetic coefficients be used as input for the model. Genetic coefficients used are indicated in (Table 4.2). The thermal time for these were calculated using the weather data collected during the field trials, which

Table 4.2 Genetic coefficients of the CERES-Maize model

Genetic coefficient	Description
P1	Thermal time from seedling emergence to the end of the juvenile phase (expressed in degree days above a base temperature of 8 °C) during which the plant is not responsive to changes in photoperiod.
P2	Extent to which development (expressed as days) is delayed for each hour increase in photoperiod above the longest photoperiod at which development proceeds at a maximum rate (which is considered to be 12.5 h).
P5	Thermal time from silking to physiological maturity (expressed in degree days above a base temperature of 8 °C).
G2	Maximum possible number of kernels per plant.
G3	Kernel filling rate during the linear grain filling stage and under optimum conditions (mg d^{-1}).
PHINT	Phyllochron interval; the interval in thermal time (degree days) between successive leaf tip appearances.

Table 4.3 Set of genetic coefficients derived for the three maize varieties

Genetic coefficient	Abontem	Omankwa	Obatanpa
P1	217.2	239.6	266.9
P5	668.8	723.8	910
G2	585	588	591
G3	8.5	8	7.5
PHINT	45.3	45	45

Note. PHINT, phyllochron interval.

were carried out under optimal soil fertility and water conditions. The phenology was calibrated using P1 and P5. The phyllochron interval (PHINT) was used together with P1 and P5 to calibrate leaf area index (LAI) and total biomass; G2 and G3 were used to partition resources into grain. The calibration data were used to derive these genetic coefficients.

With this done, the maize model was executed for each variety and the simulated yield and phenological stages were compared with data observed for the four independent trials, which were conducted under nonlimiting resource conditions. The genetic coefficients were slightly adjusted until there was good agreement between the simulated and the observed. The maize model was then considered duly calibrated for each of the three varieties. The final values of the accepted genetic coefficients are summarized in Table 4.3.

Evaluation of Model Performance

The performance of the model in predicting maize crop phenology, maximum LAI, biomass, and grain yield were assessed using a set of statistics. Data for the evaluation trials were used to test the performance of the model. The Root Mean Square Error (RMSE), coefficient of model efficiency, Willmott et al. (1985) *d* values, and the coefficient of determination were used, where RMSE is defined as:

$$\text{RMSE} = \left[\frac{1}{n}\Sigma\left(\text{yield}_{\text{simulated}_i} - \text{yield}_{\text{observed}_i}\right)^2\right]^{0.5}$$

The lower the RMSE, the better the model performance and its minimum value of 0 implies a perfect model performance. The Relative Root Mean Square Error (RRMSE) is defined as:

$$\text{RRMSE} = \left(\frac{\text{RMSE}}{\text{Observed mean}}\right) \times 100$$

The RRMSE values range between 0 and 100%. Values closer to 0 are preferred. The modified model efficiency (E_1) performance was also assessed using the equation below:

$$E_1 = 1 - \frac{\sum_{i=1}^{n} \left|\text{Observed}_i - \text{Simulated}_i\right|}{\sum_{i=i}^{n} \left|\text{Observed}_i - \text{Mean}_{\text{observed}}\right|}$$

A model efficiency of 1 denotes a perfect fit of observed and simulated data, whereas $E_1 = 0$ implies poor agreement between observed and simulated data. In the modified version, squared difference terms are replaced by their respective absolute values, thus minimizing the sensitivity of the coefficient to outliers, as indicated in the original coefficient (Legates & McCabe, 1999).

Derivation of Maize Yield Maps for the Three Northern Regions

Data from 34 soil profiles located at different sites in the five northern regions were used in this study (Fig. 4.1). Some of these sites lie in the Sudan Savanna zone (Navrongo and Bawku), which receive somewhat less rainfall than the Guinea Savanna. The soil profile data were converted to DSSAT format. Out of the 34 sites, measured weather data were available from the Ghana Meteorological Agency for only 12. For the remaining locations, the weather data were obtained from NASA POWER. The duration of the weather records ranged between 19 and 30 yr. Using the weather and soil information, the long-term yields of the three selected maize varieties were simulated. For each variety, three planting windows were considered, namely early (15 April–15 May), mid-planting (16 May–15 June), and late planting (16 June–15 July). Within each window, planting was implemented in the model when the soil water content was between 40 and 100% of field capacity within the first 30-cm depth. Each year's simulation was independent of the other years, to enable the assessment of the variability of climate on yields.

The yearly simulated maize yields (19–30 yr) for each variety at each of the 35 soil profile locations were used to generate the median yields, which were assumed to represent the average yield for a given variety per location. Using geographical information systems (GIS) software, maize yield maps were derived for the entire Guinea Savanna zone of Ghana.

Results and Discussions

Calibration and Validation of the DSSAT Model for Maize Varieties

The development and growth of the three maize varieties could generally be well calibrated with Wilmott *d* values ranging from .79 to .98 for grain yield and .73 to .97 for biomass simulation. The number of days to flowering and maturity were predicted with RMSE of between 2.5 and 3.3 d. Therefore, the set of genetic coefficients derived for each of the three varieties (Table 4.3) were considered as adequate and the model as reliable for use in further simulation studies.

The data from the evaluation trials served to further assess the model performance. The observed average maize grain yield for the extra early variety (Abontem) ranged from 683 kg ha^{-1} for the control to 3,690 kg ha^{-1} when 120 kg N ha^{-1} was applied, whereas that for the early maturity group ranged from 888 kg ha^{-1} when no fertilizer was applied to 4,086 kg ha^{-1} when 120 kg N ha^{-1} was applied. The intermediate variety, Obatanpa, had grain yields ranging from 1,030 kg ha^{-1} under the control to 4,799 kg ha^{-1} when 120 kg N ha^{-1} was applied. Biomass yields for the extra early variety ranged from 1,344 to 11,333 kg ha^{-1} for the control and 120 kg N ha^{-1} application rate, respectively. For the early maturity variety, biomass yield ranged from 1,544 kg ha^{-1} when no fertilizer was applied to 11,048 kg ha^{-1} when 120 kg N ha^{-1} was applied, and the corresponding biomass yields for the intermediate maturity variety were 2,722 and 13,600 kg ha^{-1}.

The calibrated DSSAT–CERES maize model generally predicted phenology well. Anthesis was predicted with a RMSE of 3.6 d and maturity was predicted with 5.4 d. However, the model could not capture the observed delay in maize phenology under N stress conditions. This weakness of the model in adequately simulating variations in maize life cycle as a result of N stress had earlier been reported by other studies (MacCarthy et al., 2010; Gungula et al., 2003). This is a research gap that must be addressed since maize is still grown under low soil N conditions in many tropical countries. However, the RRMSE across the three varieties was less than 15%. Other studies reported RRMSE of 18%, which was deemed as an acceptable prediction for phenology (Tesfaye et al., 2015).

Biomass yield in response to inorganic N fertilization was generally adequately predicted by the model with simulated mean values generally within the standard deviations of the average observed values (Figure 4.2). The model efficiency coefficient values ranged from .71 to .83 across varieties when comparing observed and simulated biomass yield. Coefficients of determination were above .83 and RMSE ranged between 1,265 and 1,644 kg ha^{-1}, representing RRMSE values of between 16 and 21%.

Grain yield predictions in response to the various levels of inorganic N fertilizer applications were generally well predicted for all three cultivars (Figure 4.3). The average simulated grain yield was within the standard deviation of the mean observed grain yields. The model, however, underestimated grain yield when no fertilizer was applied (Figure 4.3). The performance of the model in simulating grain yield response to inorganic N fertilizer was well simulated with E_1 values ranging from .68 to .88 across the three varieties. Grain yield of the three varieties were adequately simulated with RMSE values ranging between 454 and 583 kg ha^{-1}, representing RRMSE values ranging from 14 to 23%.

Simulation of Maize Yields in the Guinea Savanna Zone of Ghana under Diverse Nutrient, Soil Water, and Management Conditions

Given that the DSSAT model could be successfully validated for the three maize varieties, it is desirable to use the model to assess the overall impacts of factors such as soil water, N application, and plant population on yields as these are

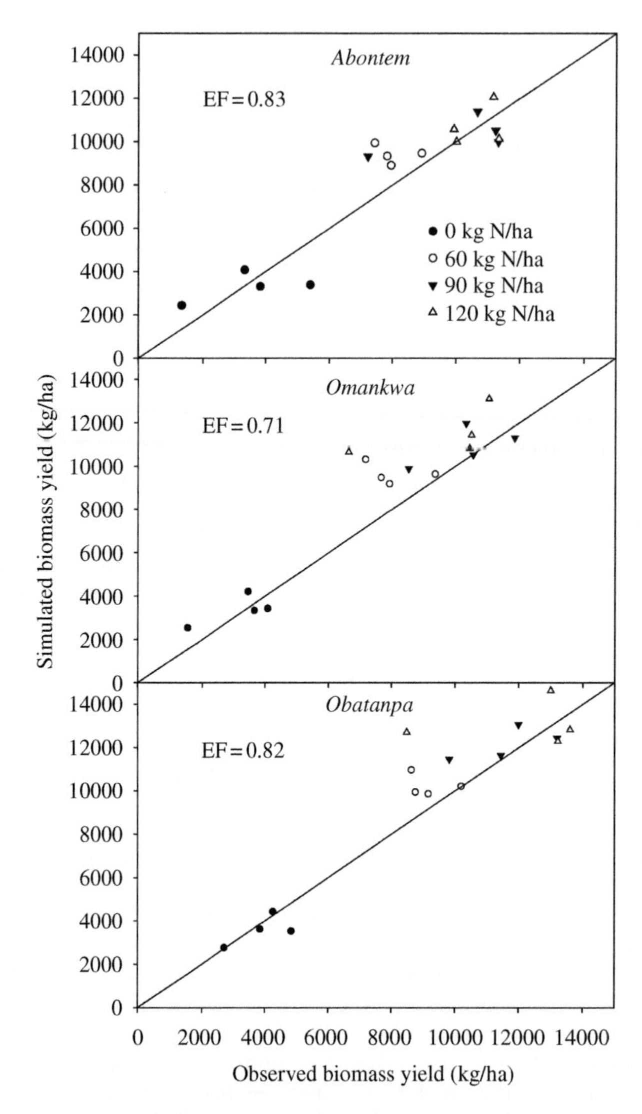

Figure 4.2 Comparison of observed maize biomass yield response to inorganic N fertilizer rate with simulated average yield response in the Northern Region of Ghana. *Abontem*, *Omankwa* and Obatanpa are extra-early, early and intermediate maturity varieties, respectively.

factors that often vary in small-holder, rain-fed systems for a year with the lowest yield in the seasonal yield simulations. Figure 4.4 illustrates the effect of N fertilizer application on leaf expansion, which is a critical physiological process for growth. The stress factor is expressed on a scale of 0 to 1, with 0 implying no stress and 1 implying maximum stress. For the control (when no mineral fertilizer was applied), N stress factor on leaf expansion remained high over much of the

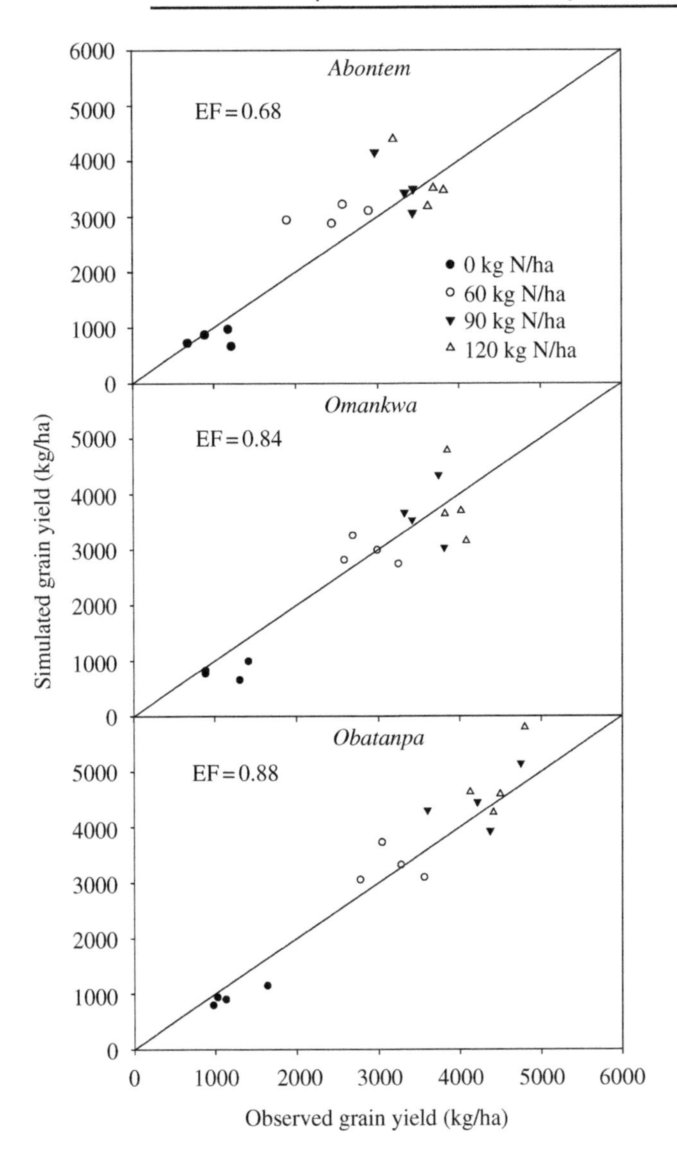

Figure 4.3 Comparison of observed maize grain yield response to inorganic N fertilizer rate with simulated average yield response in the Northern Region of Ghana. *Abontem*, *Omankwa* and Obatanpa are extra-early, early and intermediate maturity varieties, respectively.

growing season across the three varieties. The stress factors for the shorter maturity varieties (Abontem and Omankwa) were marginally lower than that of the intermediate variety (Obatanpa). The N stress factor on leaf expansion reduced with increased fertilizer use with the treatment receiving 120 kg ha^{-1} N not showing any N stress.

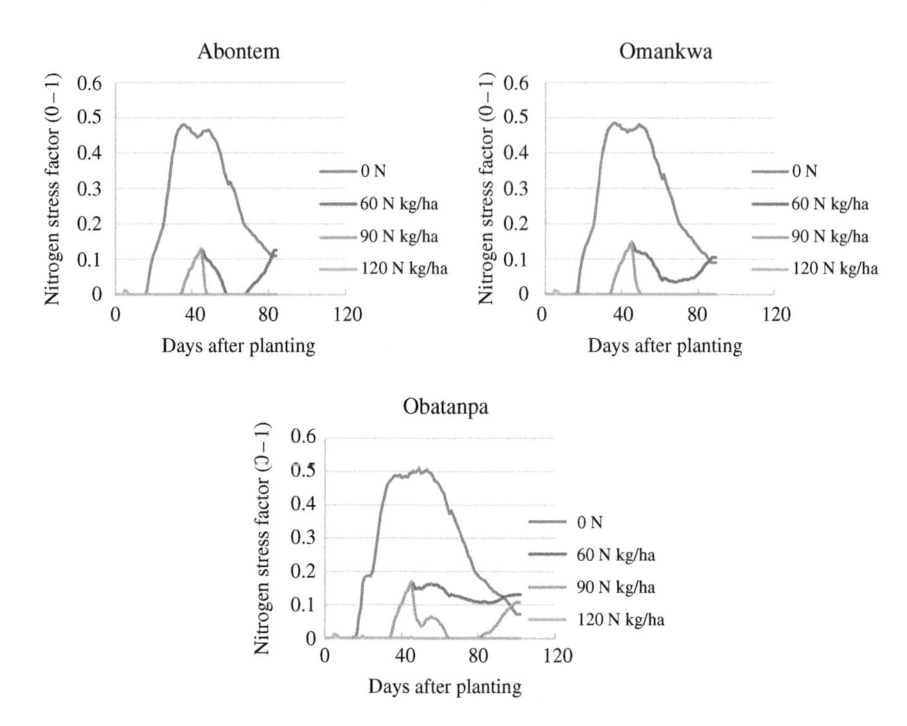

Figure 4.4 Simulated N stress factors for leaf expansion of three maize varieties under varied N fertilization rates for Tamale, Ghana, in the 2015 growing season (1 implies maximum stress and 0 implies no stress).

The DSSAT simulations showed that the addition of N fertilizer generally increased maize yields for all the years across sites (Figure 4.5). The inter-seasonal variability in yields was apparent at all levels of fertilization across the study sites. As the amount of N applied increased, the variability in maize yield also increased. For instance, for Abontem planted in the early planting window, coefficient of variation (CV) across the landscape was between 10 and 37% when 90 kg N ha^{-1} was applied, whereas the CV ranged between 10 and 34 under 60 kg N ha^{-1} application (not shown). In general, median maize yield in response to N fertilization varied from 204 kg ha^{-1} when no fertilizer was applied to 2,019 kg ha^{-1} with the application of 120 kg N ha^{-1}. The response to N fertilization was hyperbolic with no appreciable yield increases after 90 kg N ha^{-1} application (Figure 4.3). Similar trends were observed across the three other sites.

It is important to note that the application of fertilizer at any given rate and planting time does not lead to a single maize yield but to a range of yields determined largely by the season's weather conditions. Maize yield can range from as low as 3,000 kg ha^{-1} to as high as 6,000 kg ha^{-1} for the same fertilizer application rate of 120 kg N ha^{-1}. In the case of the lower yield, the farmer's investment expectation would not be met.

Figure 4.6 illustrates water stress factors under the various N treatments for each of the three varieties over the growing season. Except for the period from 21

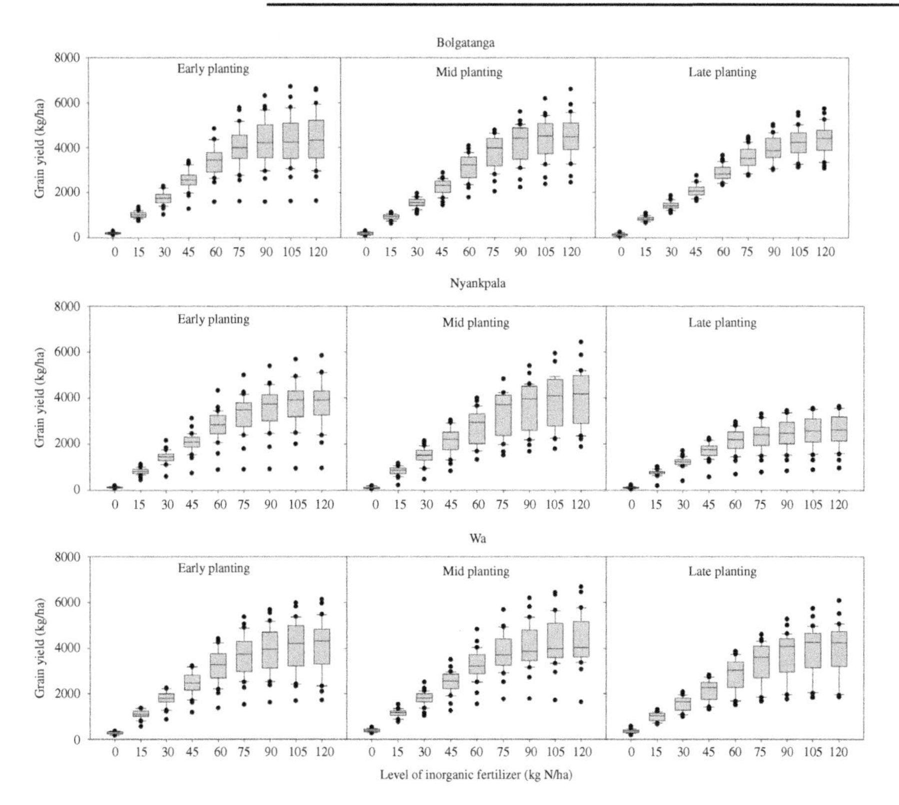

Figure 4.5 Simulated maize (Obatanpa) variability in response to N fertilizer for 30 yr for three locations in the study area over three different planting times over the growing seasons. Each box in the graph shows the distribution of grain yield over the 30-yr simulation period. The boundary of the box closest to zero indicates the 25th percentile, a line within the box marks the mean, and the upper boundary of the box indicates the 75th percentile. Whiskers above and below the box indicate the 95th and 5th percentiles, respectively.

to 29 d after planting, when high water stress of about .6 was simulated, the water stress at other times of the crops cycle was low. This suggested that the seasonal rainfall was quite well distributed and adequate for crop growth in general but also during the critical reproductive stage. Hence, water stress will not be a critical yield determinant for maize in the season. During the short period of time that water stress was high, maize plants fertilized with 120 kg N ha^{-1} experienced the highest stress across varieties. Presumably, higher N application induced higher biomass production leading also to higher use and hence a faster depletion of the soil water. But this effect, as indicated, was short-lived.

Plant population or density is one management factor that the farmer can employ to optimize resource use toward optimizing plant performance. One advantage of using simulation models is that several such parameters can be easily varied, and their impacts assessed, without necessarily embarking on tedious field experimentation. Increased plant population under no N input or low fertilization generally results in similar yields (Figure 4.7). Increased plant population is probably not an

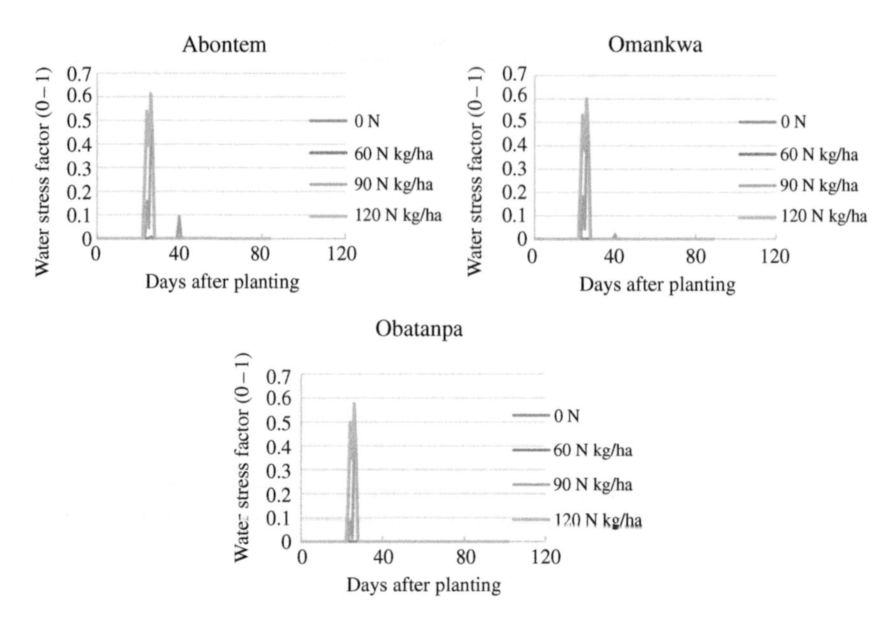

Figure 4.6 Simulated water stress factors of three maize varieties under varied N fertilization rates for Tamale, Ghana, in the 2015 growing season.

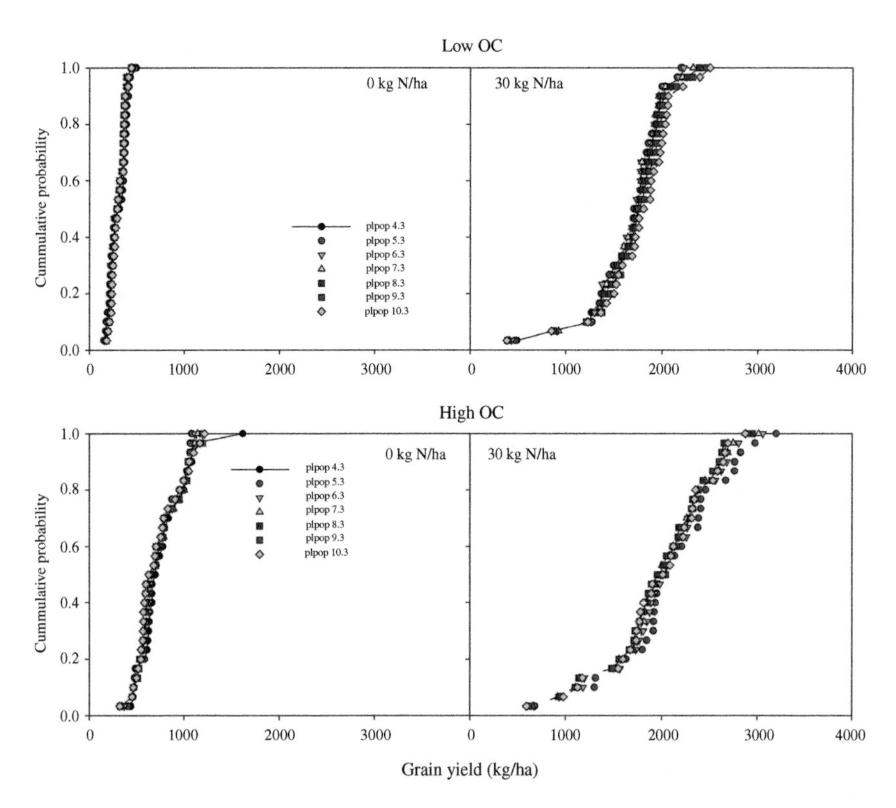

Figure 4.7 Simulated effects of plant population on the yield of maize grown under low N fertilization and soils with different fertility.

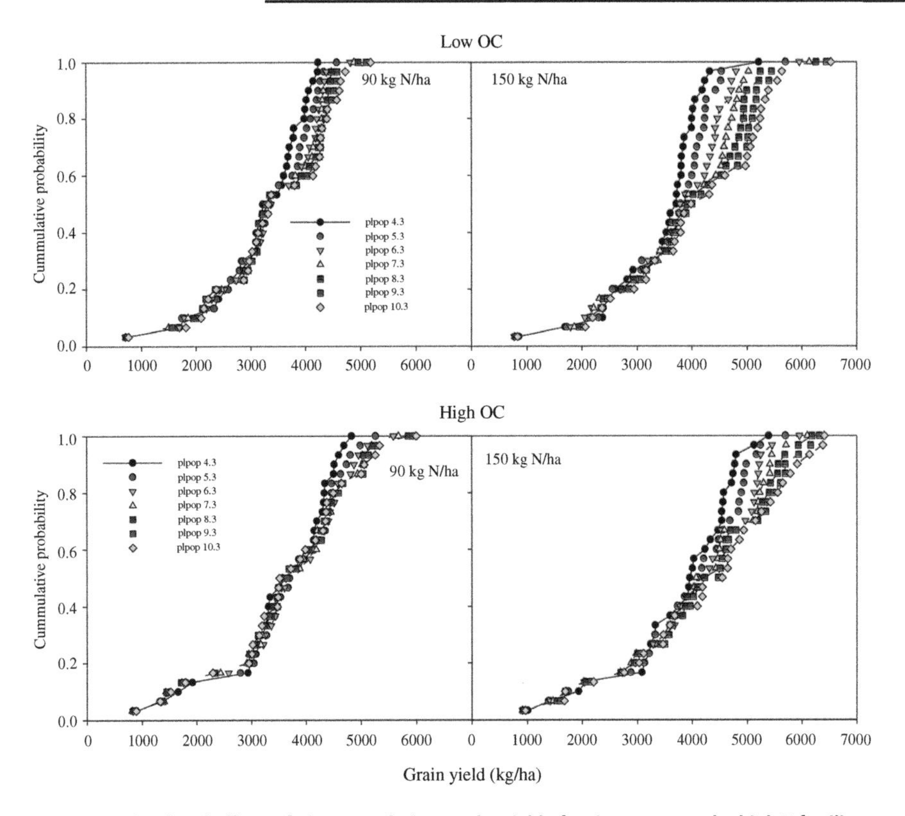

Figure 4.8 Simulated effects of plant population on the yield of maize grown under high N fertilization and soils with different fertility.

appropriate recommendation to smallholder farmers who apply little fertilizer of about 30 kg N ha^{-1} (MacCarthy et al., 2017) or less (Figure 4.7). Figure 4.8 illustrates the impact of increasing plant population on the yield of maize on soils with different soil organic C and different levels of N fertilization. Variability in grain yield generally increased marginally under the higher fertilizer level when population increases. Grain yield increase in response to increased plant population was higher under the 150 kg N ha^{-1} level whereas increased plant population resulted in similar yields on the soil with low soil organic C. Generally, curves to the right are more preferred compared with those to the left. It remains to be seen if the increases in grain yield due to increased plant population makes economic sense.

Yield Mapping and Production Domains within Northern Ghana for Various Maize Varieties

An important basis for policy and agricultural planning is the ability to fit crop varieties into their niches for optimum performance. This often requires multi-location trials for various varieties. Here, we employ the model output to derive yield maps under varied fertilizer application rates and also for different planting dates. Given that increased fertilizer use is a policy promoted by governments, we restricted our derivations to two relatively higher N application rates (60 kg N ha^{-1} [moderate] and 90 kg N ha^{-1} [high]) and ignored the lower rates.

Figures 4.9–4.11 indicate that generally a medium-time planting between 16 May and 15 June is more favorable for the Sudan Savanna zone (e.g., Bawku and Navrongo sites) than an early planting time between 15 April and 15 May or late planting time from 16 June to 15 July. This can probably be explained by the fact that the rains in the Sudan Savanna zone start later than those in the Guinea Savanna (Tamale) and cease earlier than in the Guinea Savanna. Therefore, the length of the growing season in the Sudan Savanna is shorter than in the Guinea Savanna. For the intermediate maturity variety, the late planting between 16 June and 15 July produced the least yield compared with the other two planting windows. Comparison of the yield maps (Figures 4.9–4.14) show that higher yields were obtained under 90 kg N ha^{-1} than under 60 kg N ha^{-1} for the intermediate variety, whereas differences between two N levels for the shorter maturity varieties were small.

The average grain yield response to N applied across the study site (Figure 4.15) indicates 60 kg N ha^{-1} as an appropriate or optimal N for the shorter maturity varieties (Abontem and Omankwa). The additional yield benefit may not make economic reason to invest the additional 30 kg N ha^{-1}. The intermediate variety, however, responded to the application of 90 kg N ha^{-1} compared with the 60 kg N ha^{-1}. The equations given on the curves in Figure 4.15 provide an opportunity to estimate the expected yields for a given planting date and N application rates for three varieties. In general, there is the need for economic analysis to determine which level of fertilizer input would be the best strategic investment. Under increasing climate change

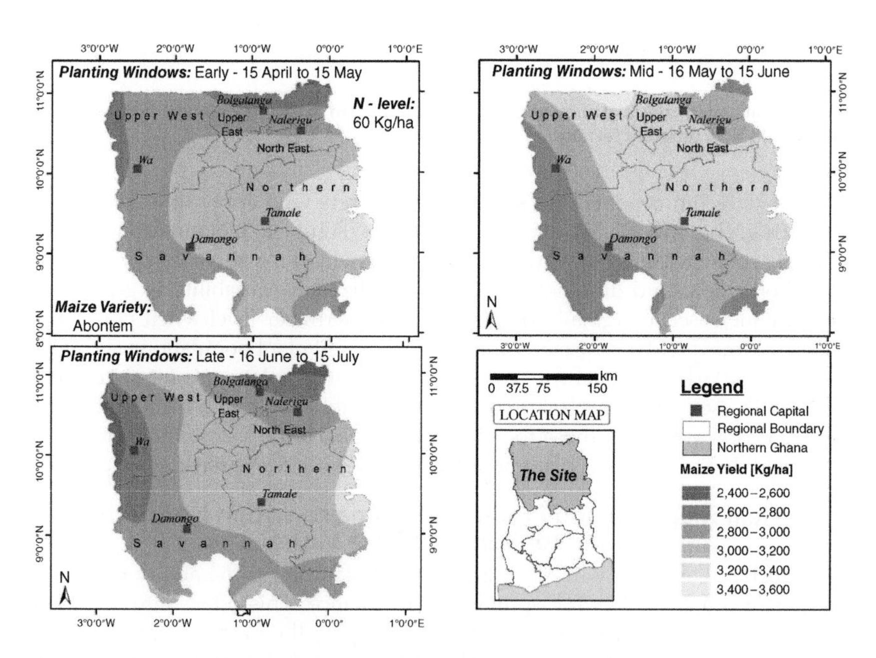

Figure 4.9 Average yield map of the Abontem (extra early maturity) maize variety at 60 kg N ha^{-1} in the northern regions over different planting windows.

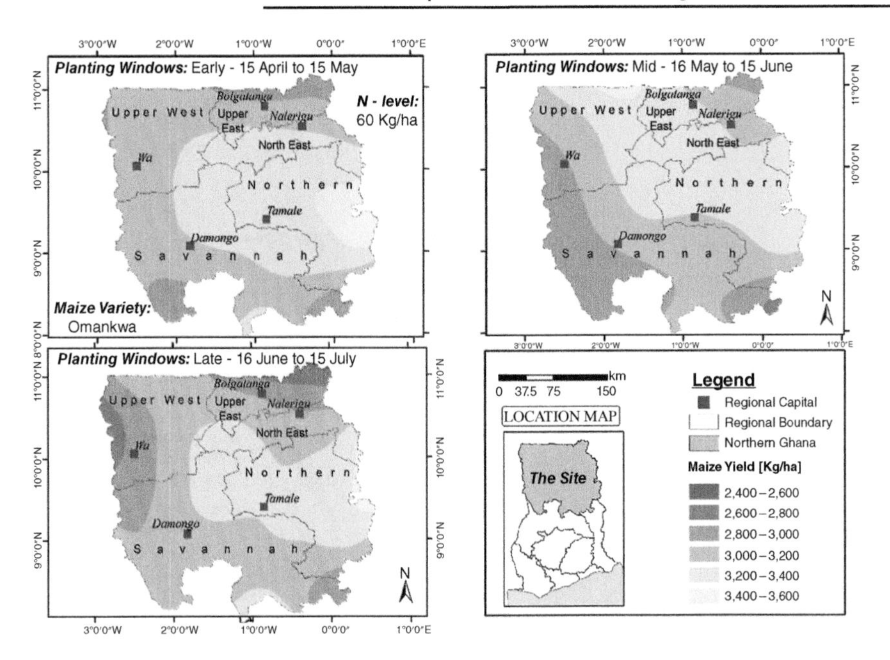

Figure 4.10 Average yield map of Omankwa (early maturity) maize varieties at 60 kg N ha^{-1} in the northern regions over different planting windows.

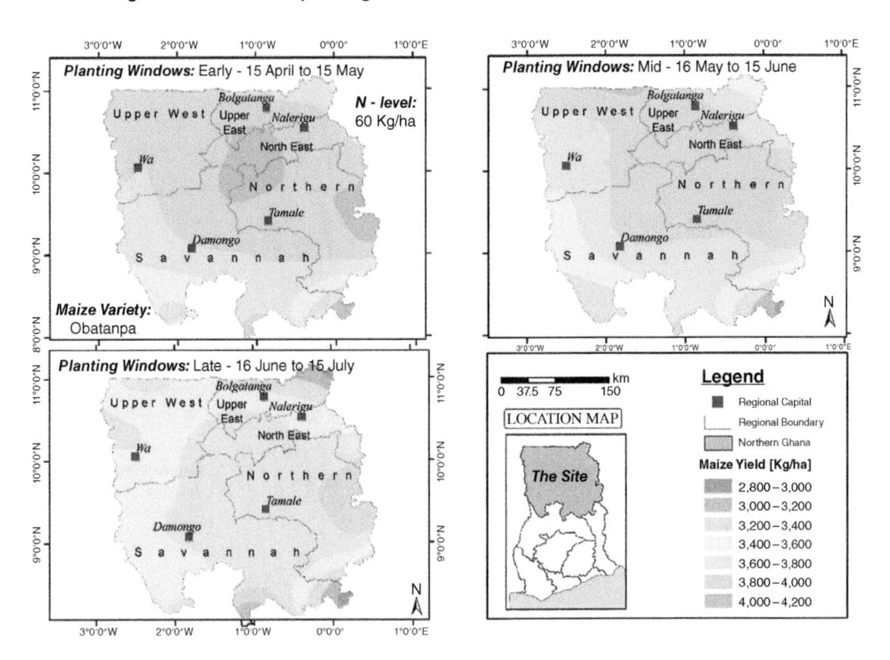

Figure 4.11 Average yield map of the Obatanpa (intermediate maturity) maize variety at 60 kg N ha^{-1} in the northern regions of Ghana at different planting windows.

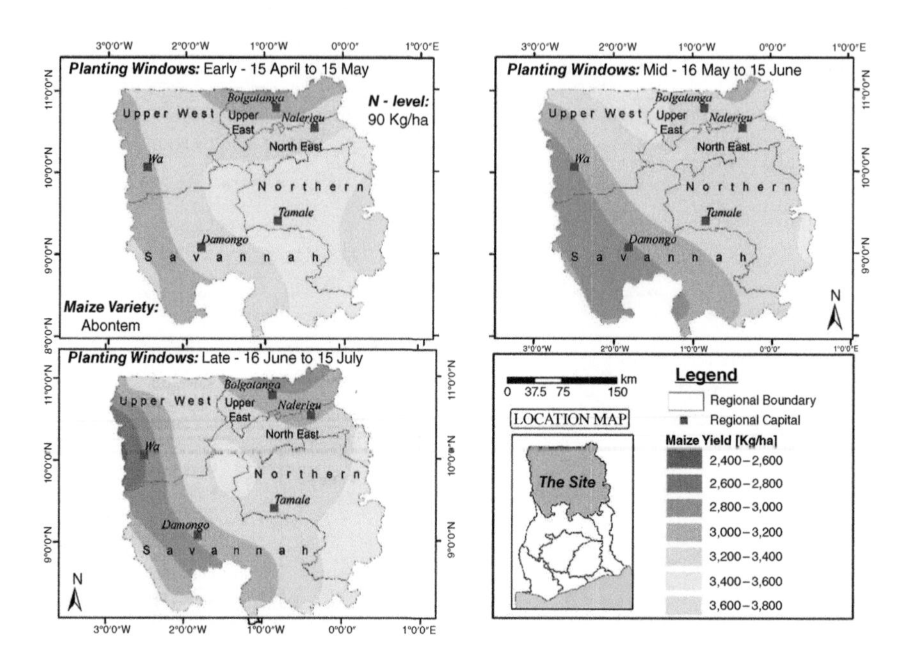

Figure 4.12 Average yield map of the Abontem (extra early maturity) maize variety at 90 kg N ha^{-1} in the northern regions over different planting windows.

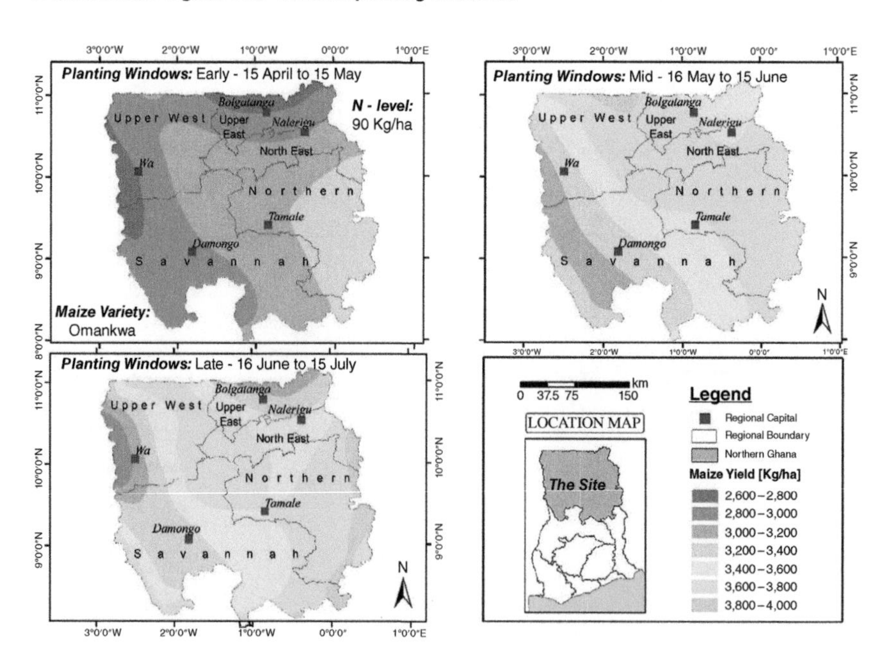

Figure 4.13 Average yield map of the Omankwa (early maturity) maize variety at 90 kg N ha^{-1} in the northern regions over different planting windows.

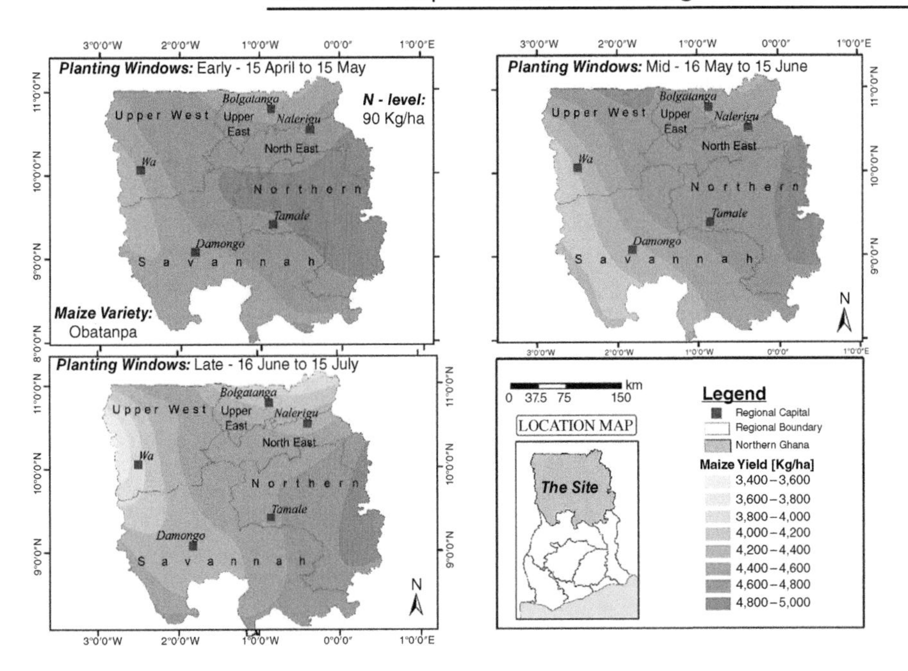

Figure 4.14 Average yield map of the Obatanpa (intermediate maturity) maize variety at 90 kg N ha^{-1} in the northern regions of Ghana at different planting windows.

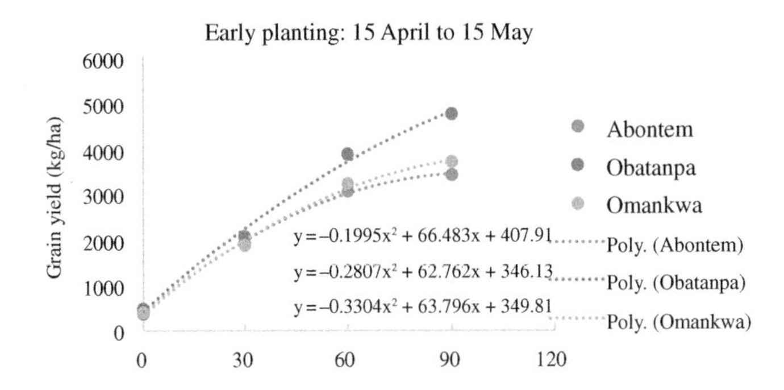

Figure 4.15 Simulated mean N response curves for the three maize varieties and three planting dates in the three regions in northern Ghana.

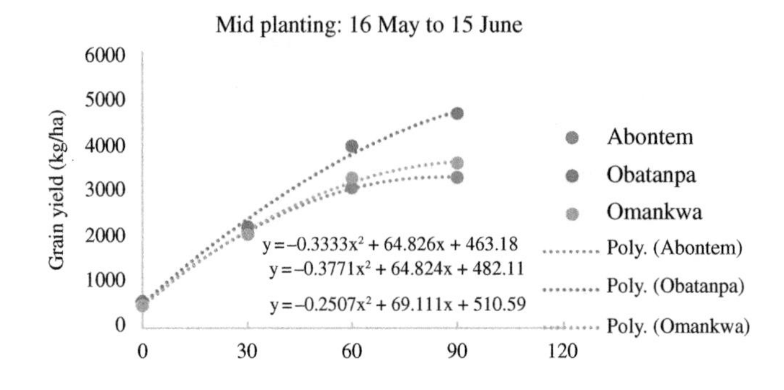

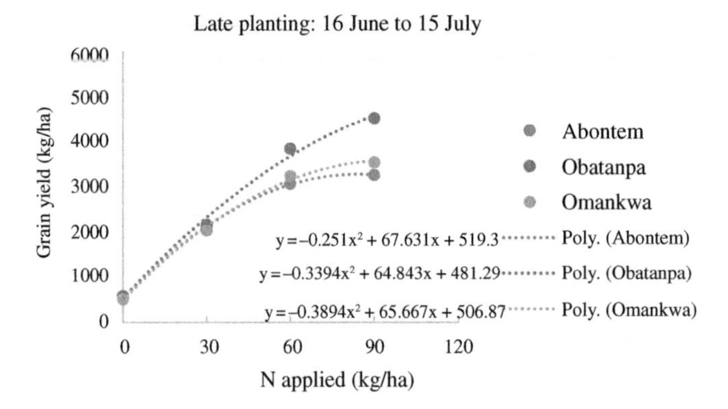

Figure 4.15 (Continued)

and variability, the question of interest would be the probability of attaining the different yield levels for a given fertilizer application rate. In essence, what soil and crop husbandry strategies should be in place to minimize the variability in yields? These questions require further research attention.

Generally, the variability in yield differs with planting time. The first planting window recorded the highest variation compared with the mid-planting and even the late planting across the study sites and varieties. In effect, in the Sudan Savanna areas (Navrongo and Bawku), planting of the three maize varieties would be recommended between 16 May and 16 June. In the southeastern part of the study area (Tamale, Walewale, and Yendi environs), early and late planting (15 April–15 May and 16 June–15 July, respectively) are preferred.

Conclusion

This study sought to employ crop simulation techniques to evaluate the performance of three different maturity maize varieties in the northern regions of Ghana. To achieve this, field experiments were conducted at several locations to calibrate and validate the DSSAT maize model for three varieties under varied soil, water, and N conditions for 2 yr. The comparison between observed and simulated yields showed good agreement.

The calibrated maize models were employed to assess the effect of varying N application on the performance of the maize varieties. The results indicated that 60 kg N ha^{-1} was adequate to provide optimal yield for the extra early and early varieties, and the intermediate varieties could still yield significant amounts of grain at 90 N kg ha^{-1} compared with 60 kg N ha^{-1}. The model was also employed to simulate long-term historical yield for the maize varieties (groups into three maturity groups) over periods of 19 to 30 yr for different planting dates for 34 profile points located across the three northern regions of Ghana. For a given planting date and N application rate, the median simulated maize yield was used to derive yield maps using GIS techniques. The long-term seasonal simulation indicates the intermediate variety with higher yield potential could still yield higher than the extra early and early varieties across the study sites. It could be shown that for Sudan Savanna area (Navrongo and Bawku), planting between 16 May and 15 June will be preferred for all three maize varieties. In the eastern part of the Guinea Savanna, early and late planting (15 April–15 May and 16 June–15 July) will be preferred than mid-planting (16 May–15 June). In general, it is concluded that crop modeling provided an effective tool that could be used to provide a sound basis for agricultural advice and policy formulation.

Acknowledgment

The authors are grateful for the International Institute for Tropical Agriculture for providing the grant (Grant no. 2100155022217, Project ID no. P-Z1-AAZ-1010 from the African Development Funds) for this study.

Abbreviations

CV, coefficient of variation; DSSAT, Decision Support Systems for Agro-Technological Transfer; CERES, Crop Environment REsource Synthesis; GIS, geographical information systems; LAI, leaf area index; PHINT, phyllochron interval; RMSE, Root Mean Square Error; RRMSE, Relative Root Mean Square Error; SARI, Savannah Agricultural Research Institute.

References

Dzotsi, K. A., Jones, J. W., Adiku, S. G. K., Naab, J. B., Singh, U., Porter, C. H., & Gijsmana, A. J. (2010). Modelling soil and plant phosphorus within DSSAT. *Ecological Modelling, 221*, 2839–2849. https:doi.org/10.1016/j.ecolmodel.2010.08.023

Fosu, M., Buah, S. S., Kanton, R. A. L., & Agyare, W. A. (2012). Modelling maize response to mineral fertilizer on siltyclayloam in the northern savanna of Ghana using DSSAT model. In J. Kihara, D. Fatondji, J. W. Jones, G. Hoogenboom, R. Tabo, & A. Bationo (Eds.), *Improving soil fertility recommendations in Africa using the Decision Support Systems for Agro-Technology Transfer (DSSAT)* (pp. 157–168). Springer Science + Business Media, B.V.

Gungula, D. T., Kling, J. G., & Togun, A. O. (2003). CERES-maize predictions of maize phenology under nitrogen-stressed conditions in Nigeria. *Agronomy Journal, 95*, 892–899. https://doi.org/10.2134/agronj2003.8920

Jones, J. W., Hoogenboom, G., Porter, C. H., Boote, K. J., Batchelor, W. D., Hunt, L. A., Wilkens, P. W., Singh, U., Gijsman, A. J., & Ritchie, J. T. (2003). The DSSAT cropping system model. *European Journal of Agronomy, 18*, 235–265.

Kugbe, J., Fosu, M., & Vlek, P. L. (2015). Impact of season, fuel load and vegetation cover on fire mediated nutrient losses across savanna agroecosystems: The case of northern Ghana. *Nutrient Cycling in Agroecosystems, 102*, 113–136.

Legates, D. R., & McCabe, G. J. (1999). Evaluating the use of 'goodness-of-fit' measures in hydrologic and hydro-climatic model validation. *Water Resources Research, 35*, 233–241.

MacCarthy, D. S., Adiku, S. G. K., Freduah, B. S., Gbefo, F., & Kamara, A. Y. (2017). Using CERES-Maize and ENSO as decision support tools to evaluate climate-sensitive farm management practices for maize production in the northern regions of Ghana. *Frontiers in Plant Science, 8*, 31. https://doi.org/10.3389/fpls.2017.00031

MacCarthy, D. S., Agyare, W. A., & Vlek, P. L. G. (2018). Evaluation of soil properties of the Sudan Savannah ecological zone of Ghana for crop production. *Ghana Journal of Agricultural Science, 52*, 95–104.

MacCarthy, D. S., Vlek, P. L. G., Bationo, A., Tabo, R., & Fosu, M. (2010). Modeling nutrient and water productivity of sorghum in smallholder farming systems in a semi-arid region of Ghana. *Field Crops Research, 118*, 251–258.

Ritchie, J. T. (1998). Soil water balance and plant water stress. In G. Y. Tsuji, G. Hoogenboom, & P. K. Thornton (Eds.), *Understanding options for agricultural production* (pp. 41–54). Kluwer Academic Publishers.

Saxton, K. E., & Rawls, W. J. (2006). Soil water characteristic estimates by texture and organic matter for hydrologic solutions. *Soil Science Society of America Journal, 70*, 1569–1578. https://doi.org/10.2136/sssaj2005.0117

Tesfaye, K., Gbegbelegbe, S., Cairns, J. E., Shiferaw, B., Prasanna, B. M., Sonder, K., Boote, K., Makumbi, D., & Robertson, R. (2015). Maize systems under climate change in sub-Saharan Africa. *International Journal of Climate Change Strategies and Management, 7*, 247–271.

Willmott, C. J., Ackleson, G. S., Davis, R. E., Feddema, J. J., Klink, K. M., Legates, D. R., O'Donnell, J., & Rowe, C. M. (1985). Statistics for the evaluation and comparison of models. *Journal of Geophysical Research, 90*, 8995–9005. https://doi.org/10.1029/JC090iC05p08995

5

Shakir Ali,
Adlul Islam,
and Prabhat R. Ojasvi

Modeling Water Dynamics for Assessing and Managing Ecosystem Services in India

Abstract

The agro-ecosystem is one of the world's most important ecosystems, which provides food, fiber, fodder, and fuel, as well as water and climate regulation, and cultural services to humans for their well-being. These agro-ecosystem services also provide significant economic benefits to people and the socio-economic development of a country. However, the sustainability of agro-ecosystem services is threatened by climate change and variability. Agro-ecosystem models have been recognized as a potential tool to understand the interaction between climate and other components of an agro-ecosystem and identify the best management practices for sustainable management of the system. This chapter presents the application of the agro-ecosystem models to simulate some of the water-regulating agro-ecosystem services under climate change scenarios in India. Simulation models have been successfully applied for the infiltration, groundwater recharge, evapotranspiration, evaporation, runoff, water storage, and soil erosion as the water-regulating agro-ecosystem services. The application of various infiltration models developed in India has effectively been demonstrated for both constant and variable depth of water for a varicty of soils. A large variation in potential groundwater recharge both in time and space has been recorded for the soybean [*Glycine max* (L.) Merr.] cropping system under different soils using the

Enhancing Agricultural Research and Precision Management for Subsistence Farming by Integrating System Models with Experiments, First Edition. Edited by Dennis J. Timlin and Saseendran S. Anapalli.

DOI: 10.1002/9780891183891.ch05

HYDRUS-1D model. Simulated potential groundwater recharge from water harvesting structures (WHS) by the integrated potential groundwater recharge simulation (IPGRS) model ranged from 83.0 to 90.2% of the accumulated surface runoff in the WHS. Different hydrological models, such as the Precipitation Runoff Modeling Systems (PRMS), and the Soil and Water Assessment Tool (SWAT) have been applied successfully to various river basins of India for modeling climate change impact on water resources. Simulation of soil erosion using the revised universal soil loss equation (RUSLE) model under HADCM3 model simulated climate change scenarios under SRES A2 emission scenarios showed an increase in soil erosion from 58 to 59% for the different land-use systems with an increase of 58% in annual rainfall during the 2080s (2071–2099). Simulation studies conducted on some water-regulating agro-ecosystem services using hydrological models with different climate change projections and downscaling approaches showed varied hydrological responses of land-use systems and river basins in agro-ecosystems of India to the future climate change scenarios. In conclusion, these agro-ecosystem modeling applications tested in the semi-arid region of India can be extended to other agro-ecological regions of the world for better understanding and quantifying water-regulating agro-ecosystem services.

Introduction

Globally agro-ecosystems are the dominant forms of land management, which cover nearly 40% of the earth's surface (Alison, 2010). They play a key role in overall socio-economic development and human well-being of the countries. Agro-ecosystems are the mainstay of the economy of the undernourished people living in developing countries. Nearly 58.4% of the Indian population depends on agro-ecosystems and the allied sectors for their livelihoods, and agro-ecosystems support the vast majority of the low income, the poor, and the vulnerable people in India (Nagamani & Mariappan, 2017). Globally, agro-ecosystems show great disparity both in structure and function, as they are represented by diverse climatic regions, cultures, and socio-economic conditions. The functioning of an agro-ecosystem is an anthology of various components, which include mono and mixed cropping systems, plantations, agro-horticulture systems, agro-forestry systems, grazing systems, pastoral systems, shifting cultivation systems, and species-rich home gardens (Alison, 2010). This variety of agricultural systems results in a highly variable assortment and quantity of ecosystem services (Alison, 2010). The agro-ecosystems in the world have undergone major structural changes due to urbanization, suburbanization, and industrialization. In India, per capita availability of arable land has dwindled from 0.48 ha in 1950 to 0.15 ha in 2000 and is likely to further reduce to 0.08 ha by 2020 (FAO, 2018). These intensification or structural changes of agro-ecosystems, including climate changes and variability, has further degraded the agro-ecosystem services (Alison, 2010; Matson et al., 1997).

Agro-ecosystem services are defined as tangible and/or intangible benefits provided from the agro-ecosystem to the people's well-being, society and economy of a country (Guswa et al., 2014; MEA, 2005). Traditionally, agro-ecosystems have been considered primarily as sources of provisioning services, the income, and the

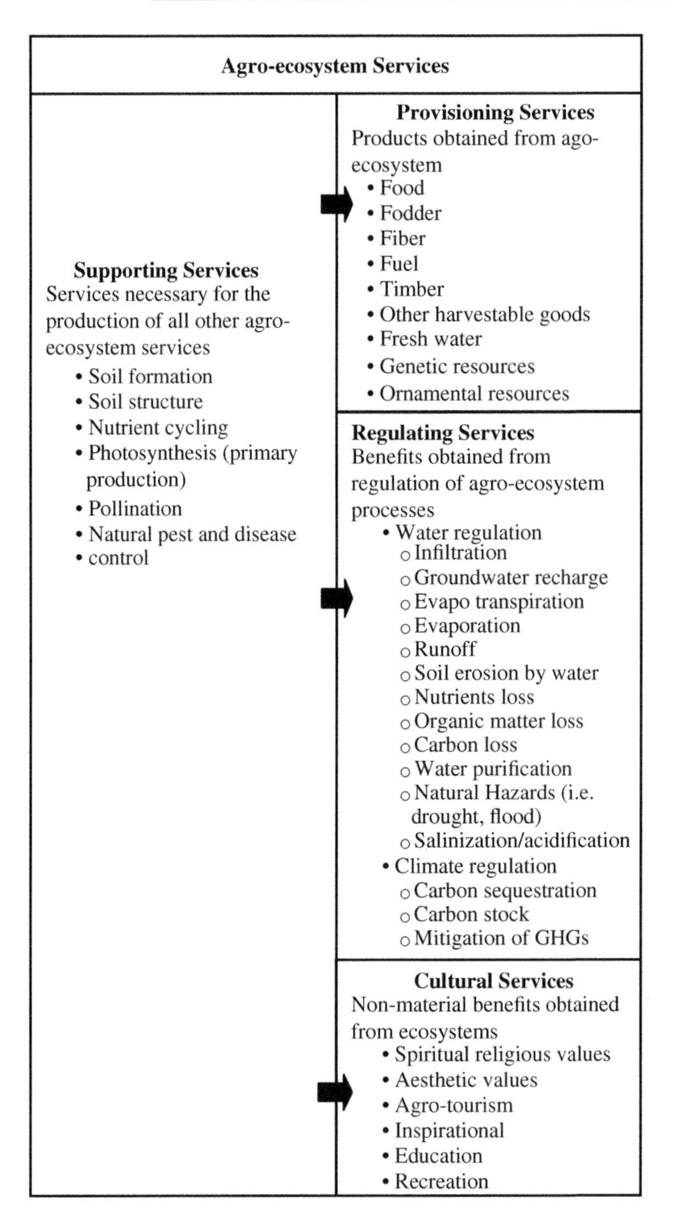

Figure 5.1 Ecosystem services provided by the agro-ecosystem.

use of products from agriculture. However, recently, a broad range of ecosystem services from agro-ecosystems have been recognized that include provisioning, regulating, supportive, and cultural and climate services (MEA, 2005) (Figure 5.1). Agro-ecosystem provisioning services include food, fodder, fiber, fuel as bio-energy, other harvestable goods, genetic resources (i.e., medicinal purposes, gene banks, fish harvested from water bodies, etc.) and ornamental resources (i.e., decorative plants) (Alison, 2010; Robertson et al., 2014). Provisioning services usually depend on regulating and supporting agro-ecosystem services. The regulation services include water regulation (i.e., hydrological flow), and climate regulation through greenhouse gas emissions. They provide a stable, healthy, and resilient environment to the people. They are slightly less tangible, and therefore this service is most difficult to assess economically (Dominati et al., 2010; Vereecken et al., 2016). Regulating services are usually overlooked. However, they are equally vital as provisioning services and they also promote agricultural provisioning services.

Water regulation services of the agro-ecosystem are linked to movement and storage of water, both in terms of quantity and quality of water. They include hydrological processes such as infiltration, evaporation, evapotranspiration (ET), groundwater recharge (GR), runoff, and associated natural hazards (i.e., flood and drought). They impact water purification, wastewater treatment, and agricultural water management practices (i.e., irrigation and drainage). They also affect land degradation processes, which include soil erosion process by water, loss of nutrients and organic matter, salinization, acidification, and biodiversity. Climate regulation is defined as the capacity of the agro-ecosystem to control states and fluxes of energy, water, and matter that impact climate, which includes greenhouse gas emissions. Supporting services refer to fundamental soil and plant production processes such as soil formation, soil structure, nutrient supply/cycling, photosynthesis (primary production), pollination, and natural pest and disease control that are essential for providing provisioning services (Chittarpur & Patil, 2017; Smukler et al., 2012). Cultural services refer to nonmaterial benefits obtained from ecosystems such as the scenic beauty, esthetics, inspirational, educational and recreational benefits, tourism, and uses in traditional rituals and customs (Daily & Matson, 2008). Conservation of biodiversity may also be considered a cultural ecosystem service influenced by agriculture.

The interactions among agro-ecosystem services are very complex as they depend on multiple and interrelated ecosystem services. To provide the primary provisioning services, intensive agro-ecosystem practices have resulted in the degradation of other valuable agro-ecosystem services such as soil fertility, soil erosion and landscape changes, water purification, climate regulation, and biodiversity conservation (Pathak et al., 2017). The Millennium Ecosystem Assessment (MEA, 2005) reported that about 60% (15 out of 24) of services measured in the assessment were being degraded or unsustainably used as a consequence of agro-ecosystem management practices and other human activities. Further, modern industrial agro-ecosystem activities have most significantly undermined the Earth life support system (Pathak et al., 2017; Rockström et al., 2004).

Water is the most essential component for ecosystem functioning (Falkenmark, 2003). Agricultural water management is the central entry point for minimizing trade-offs and finding synergies between food production and other ecosystem services (Gordon et al., 2010). Assessment of water-related ecosystem services not only helps to address water security problems but also enhances food and energy security through the management of water (Chen et al., 2018; Sahle et al., 2019). The purpose of this chapter is not to give details of the agro-ecosystem models, but to give the applications of various models to quantify and predict the water-regulating agro-ecosystem services as affected by the climate change under Indian condition. As water plays a key role in sustaining human well-being, we focus herein on water-regulating services, especially water movement and water storage in agro-ecosystems.

Modeling of Agro-Ecosystem Services

Agro-ecosystems have become more complex in recent years due to many driving forces such as growing population and its demands for ecosystem services, dwindling arable lands of agro-ecosystem for expanding provisioning services, changes in land-use systems, and increasing pressures on natural resources. The complexity of an agro-ecosystem is further compounded by climate change and variability, which creates a wide spectrum of challenges to the sustainability of agro-ecosystem services and the environment for well-being of humans and their livelihood (Jones et al., 2016; Turner et al., 2016). To overcome these changing and multiple problems of agro-ecosystems, there is growing interest in the application of agro-ecosystem models, which help basic understanding and quantification of its component(s) or interactions of the components that predict/simulate the overall performance of the agro-ecosystem. Agro-ecosystem models also help system managers and policymakers to screen the potential risk areas and identify the best management practices (BMPs) to maximize sustainability and/or profitability to maintain food and water security, and environmental quality. The uses of agro-ecosystem models have been aided in the past two decades by our ability to collect real-time dynamic field data of the soil–plant–atmosphere system using new electronic devices (technologies) such as remote sensing with satellites or proximal sensing with ground-based instruments, and availability of improved hardware and software at an affordable cost. Agro-ecosystem modeling requires a thorough understanding of the system, mathematical relationships within ecosystem components and interactions among them, real field data and expertise for their development, successful verification, calibration and validation of the model, sensitivity analysis, and prediction or simulation of the system response.

Agro-Ecosystem Models

An agro-ecosystem model is a simplified representation of a complex agro-ecosystem, which simulates the temporal or spatial response of the agro-ecosystem. Numerous agro-ecosystem models have been developed at different scales such as field, farm, regional, national, and global levels. Users of agro-ecosystem models range from farmers to policymakers interested in improving decisions and policies from field to national and global levels. The field-scale models are used to analyze responses of the system similar to experiments on real systems. A good example of a field-level model is a cropping system model, which is used to predict economic yield of the crop under different combinations of management practices and crop seasons. Field-level models usually assume spatial homogeneity across the field. A field crop model may be used to simulate multiple homogeneous fields across a farm or large landscape with input data sets of each field. The field-level models can be scaled up at the farm or regional, or even national, levels using powerful computer systems, if suitable input data are available (Elliott et al., 2014). At larger scales (i.e., farm, regional, national, and global), ecosystem models usually include biophysical responses of ecosystem as well as socio-economic, environmental, policy, and business issues. Models at each of these scales may be developed by using a field-level model. For example, a national crop model may make use of field-scale crop models to simulate production across many districts, then aggregated to the national scale for use in an economic model of the policy impacts on the aggregate production or its variations across districts.

Modeling of Water-Regulating Agro-Ecosystem Services under Climate Change

In this section, we will deal with the application of models in understanding, quantifying, and predicting agro-ecosystem regulating services. Water movement in the agro-ecosystem involves the infiltration of water into the soil and its subsequent release to the atmosphere as evaporation from soil and transpiration from the plant (i.e., ET), evaporation from the water surface, groundwater recharge, and surface runoff. Water movement as surface runoff with a velocity greater than erosive velocity causes soil erosion and loss of soil nutrients, organic matter, and C stock. We herein discuss and illustrate the basic applications of some models for several important water-regulating services.

Infiltration

Infiltration is an essential component of agro-ecosystems, which involves the entry of water from the soil surface into the soil profile, and subsequent release of this water below the root zone as deep percolation and potential recharge, and finally joining the water table as GR. The infiltration in an agro-ecosystem is controlled by rainfall intensity, soil texture, soil layers, land slope, soil moisture content, surface sealing and crusting, movement and entrapment of soil air, plant architecture (i.e., close or broad spacing), and amount of surface litter at the surface or below the surface. Agro-ecosystem management practices (i.e., tillage practices, manure, mulch, and so forth), land-use systems (agriculture, fallow, and vegetation), and soil organic C stocks significantly alter the infiltration rate. The agro-ecosystem has a higher infiltration rate than the fallow land ecosystem, and it tends to reduce peak flow and floods and increase soil moisture status and GR.

Many models have been developed and used to simulate the infiltration behavior of the upper soil layer in the agro-ecosystem at point scale. These models can be classified as empirical, semi-empirical, and physically based models. The empirical and semi-empirical models include Kostiakov (1932), Horton (1933), Philip (1957), and Holtan (1961) models, which are derived, based on field/laboratory experimental data, and used in simple mathematical equations. These models are unable to fully describe the infiltration process. However, physically based models substantially describe the infiltration process. The most commonly used process-based water flow and infiltration model is the Richards' equation (Richards, 1931). Richards' model is a combination of the Darcy equation with the continuity equation and includes a sink term for soil water extraction by roots. Richards' model is solved using an iterative implicit numerical technique with fine discretization in time and space. The Green–Ampt (GA) model (Green & Ampt, 1911) is a simplified version of Richards' equation for infiltration. This model and its several modifications have been widely used for simulating one-dimensional vertical infiltration into the soil due to its simplicity (Vatankhah, 2015; Ali et al., 2016a). These process-based models are still not suitable for all soil types and conditions, such as some soils with high clay or organic matter contents and macropores (Beven & Germann, 2013).

Very recently, following the work of Ali et al. (2013) and Ali and Ghosh (2016), Ali and Islam (2018) developed an accurate explicit model to replace the implicit GA infiltration equation using a two-step curve-fitting approach. The developed model compared well with the implicit GA model with a maximum percent relative error (MPRE) of .146 and .012% for the dimensionless cumulative and infiltration rate, respectively, and respective percent bias (PB) of .070, and .0005. Field applications of the model over a variety of soil textures (Figure 5.2) indicated its potential for application with MPRE ≤ .130% and PB ≤ .050 for cumulative infiltration and MPRE ≤ .110% and PB ≤ .080 for infiltration rate. The estimated initial and final infiltration rates of the Wadi sand and loamy sand by using the derived model were 54.2 and 39.2 cm h^{-1}, and 15.9 and 3.7 cm h^{-1}, respectively. The cumulative infiltrations were 115.5 and 15.2 cm for the Wadi sand and loamy sand, respectively, during the 2.5-h experimentation time. The derived model for infiltration rate and cumulative infiltration is defined mathematically as:

$$F(t) = \frac{s^2}{2K_s}\left\{t^* + 2.5009\ \ln\left[1 + 0.5833\sqrt{t^*}\right]\begin{bmatrix} 0.9723 + 0.0117\left[1 - \exp\left(-27.36t^*\right)\right] \\ +0.0162\left[1 - \exp\left(-2.5168t^*\right)\right]\end{bmatrix}\right\} \tag{5.1}$$

and

$$f(t) = K_s\left[1 + \frac{\eta_f(H + \psi_f)}{F(t^*)}\right] \tag{5.2}$$

where

$$t^* = \frac{K_s t}{\eta_f(H + \psi_f)} = \frac{2K_s^2 t}{s^2} \tag{5.3}$$

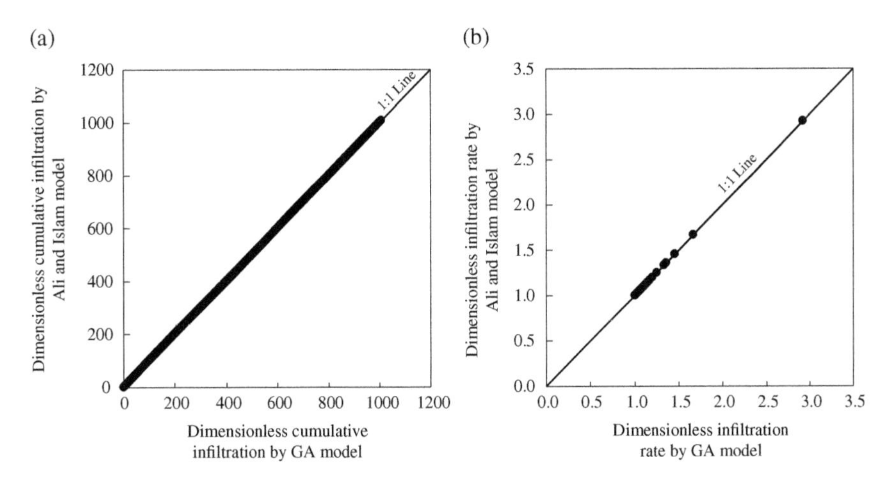

Figure 5.2 Performance of dimensionless (a) cumulative infiltration, $I^*(t^*)$, and (b) rate of infiltration, $i^*(t^*)$ estimated using the explicit model (Ali & Islam, 2018) and implicit GA model. Adopted and modified from Ali and Islam (2018).

where $F(t)$ is the cumulative infiltration at time t^* [L]; $f(t)$ is the infiltration rate at time, t [LT^{-1}]; t is the time [T]; t^* is the dimensionless time [–]; K_s is the saturated hydraulic conductivity of the transmission zone [LT^{-1}]; H is the depth of water over surface [L]; ψ_f is the suction head at the wetting front (negative pressure head) [L]; η_f is the fillable porosity [–]; and equal to $\theta_s - \theta_i$; in which θ_i is the initial volumetric soil moisture content [dimensionless]; and θ_s is the total porosity (i.e., volumetric water content at near or fully saturation) [dimensionless]; and s is the sorptivity parameter [L T$^{-1/2}$].

The infiltration studies conducted in different agro-ecosystems of India (Table 5.1) showed that the infiltration rate varied from one agro-ecological region to anther agro-ecoregion depending upon the soil and land use system. The basic infiltration rate for agriculture land use systems under clay, sandy loam, and loamy sand soils ranged from 0.2 to 1.6, 4.6 to 8.5, and 4.0 to 13.1 cm h^{-1}, respectively.

Table 5.1 Basic infiltration rate in different agro-ecoregions of India

Agro-eco regions	Locations/places	Soil types	Land use system	Basic infiltration rate cm h^{-1}	References
Assam	Dudhnai	Silty loam	Barren	0.5	Kumar et al. (1995)
		Loam	Grass	2.3	
		Silt	Agriculture	1.1	
West Bengal	Kharagpur	Sandy loam	Vegetation cum grass	1.3	Mahapatra et al. (2020)
		Sandy loam	Waste land	3.4	Machiwal et al. (2006)
Gujarat	Lalpur	Entisols	Agriculture	2.4	Gundalia (2018)
	Kankachiyala	Inceptisols		3.6	
	Chhelanka	Inceptisols		4.3	
	Vajadi	Vertisols		8.2	
J & K	Dal Lake catchment	Clay	Agriculture	1.6	Rasool et al. (2021a 2021b)
		Loamy sand		4.0	
		Sandy clay		1.1	
		Clay	Shrubland	0.8	
		Loamy sand		4.8	
		Sandy clay		5.9	
		Loamy sand	Barren land	1.3	
Madhya Pradesh	Narsinghpur	Clay	Agriculture	0.2	Roy & Singh (1995)
		Sandy clay		0.3	
		silty Clay		0.4	
Maharashtra	Sangola, Solapur	Clay	Agriculture	1.2	Dagadu & Nimbalkar (2012)
		Sandy loam		8.5	
	Kopargaon	Clay		0.2	Jejurkar & Rajurkar (2014)
Sikkim	Gangtok	Sandy loam	Agriculture	4.6	Patle et al. (2019)
		loamy sand		13.1	

Groundwater Recharge

Groundwater recharge is the process for the replenishment of groundwater storage. The GR processes have two components: one that considers advancement of wetting front by vertical movement of water through the unsaturated zone and continues until the wetting front touches the water table, known as potential infiltration/GR; another one is subsequent recharge after the wetting front touches the groundwater table, known as actual GR or GR (Ali, unpublished PhD thesis, 2009; Ali et al., 2013). The groundwater table evolves after the start of the actual GR process. The amount and timing of GR are controlled by hydrogeological and climatic conditions (Flint & Ellett, 2004). The prime factors affecting GR are rainfall, depth of water, soil type, vegetation characteristics, and so forth. According to Rushton (1997), the arrival of potential GR to the water table depends on unsaturated zone processes and the ability of the saturated zone to accept it.

A study on potential GR in Badakheda agro-ecosystem watershed in Rajasthan (India) was conducted using HYDRUS-1D. In the Badakheda watershed, the water table is declining due to increasing demands for expansion of agriculture, continuous failure of monsoon, and nonadoption of artificial GR facilities. The water table decline has resulted in drying up of dug/tube wells, lowering of water yields, and higher pumping costs. These have forced the government to execute the program for artificial replenishment of the aquifer by means of field bunding, minor leveling inter-bunded area, recharge ponds (RPs), recharge filters, and conservation trenching in the watershed to arrest the depleting water table and increase GR. Analysis of results indicated a large variation in potential recharge with both time and space, which ranged from 2.1 to 9.3, 5.4 to 17.6, and 10.2 to 23.3% of annual rainfall for the soybean [*Glycine max* (L.) Merr.] cropping system under silty clay, silty clay loam, and silty loam soils, respectively (Figure 5.3) (Khatal et al., 2018). The mean annual potential recharge for soybean cropping systems under silty clay, silty clay loam, and silty loam soils was found

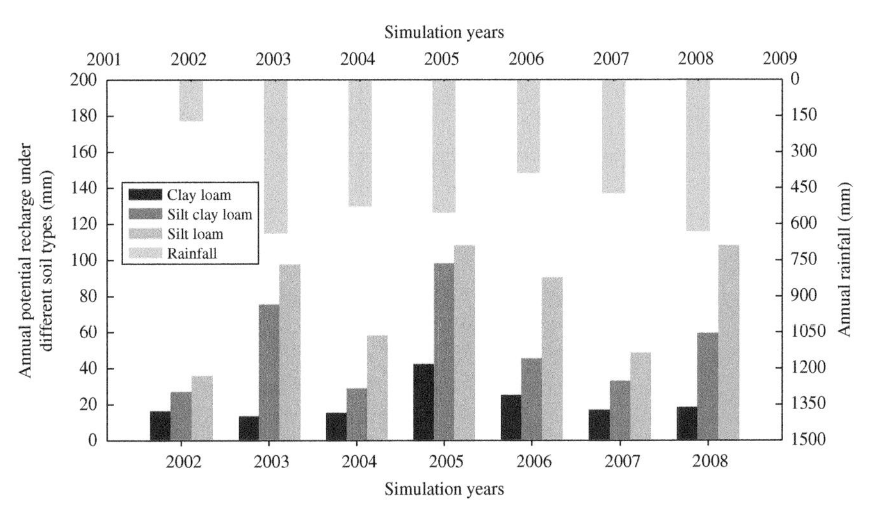

Figure 5.3 Annual potential recharge from the soybean cropping system under different soil types in Badakheda watershed, Rajasthan, India. Adopted and modified; Khatal et al. (2018).

to be 4.3, 10.8, and 16.1% of average annual rainfall, respectively. The potential recharge rates from the sparsely vegetated sand dune region and the rain-fed region of Rajasthan were found to be 2–3 and 10–16% of annual rainfall, respectively (Scanlon et al., 2010). These studies indicated that the higher clay content decreased the potential recharge and the vegetation influences recharge through transpiration and interception. The deep-rooted trees remove more water than shallow-rooted grasses.

Ali (unpublished PhD thesis, 2009) developed an integrated potential groundwater recharge simulation (IPGRS) model employing the modified Green–Ampt equation under the variable depth of water in the water balance equation. The IPGRS model offers an estimate of time-varying potential GR rate under variable water depth from the water harvesting structures in an agro-ecosystem. The IPGRS model includes time-varying rainfall, runoff, surface water evaporation, outflow, and length of the advancement of the wetting front. The model also considered saturated hydraulic conductivity and fillable porosity of the pond's bed material as their parameters. The model is process-based and holistic in nature, easy to use, and capable to simulate potential GR rate with reasonable accuracy. The model has wide field applicability and can successfully be extended for estimation of potential GR rate from the water harvesting structures of any shape and size, located in any geographical regions of India and elsewhere. The IPGRS model is defined as (Ali & Ghosh, 2019):

$$R_p(t) = \frac{K_s}{\left[L_f(t)A_{ws}(t) + K_s A_{rs}(\bar{t})\Delta t\right]} \left\{ \begin{array}{l} H(t-\Delta t)A_{ws}(t-\Delta t) + \left[Q(t)A_w + P(t)A_s - E(t)\bar{A}_{ws}(t) - Q_0(t)\right]\Delta t \\ + \left[L_f(t) + \psi_f\right]A_{ws}(t) \end{array} \right\} \quad (5.4)$$

in which the $l_f(t)$ is:

$$L_f(t) = L_f(t-\Delta t) + \left[H(t-\Delta t) + \psi\right] \left\{ \begin{array}{l} \sqrt{\dfrac{F_1 K_s}{\eta\left[H(t-\Delta t)+\psi\right]} t + F_2} \\ -\sqrt{\dfrac{F_1 K_s}{\eta\left[H(t-\Delta t)+\psi\right]} (t-\Delta t) + F_2} \end{array} \right\} \quad (5.5)$$

where $R_p(t)$ is the potential recharge rate from the RP at time t [LT^{-1}]; $Q_i(t)$ is the runoff into RP at time t [LT^{-1}]; $P_i(t)$ is the rainfall over the RP at time t[LT^{-1}]; $E_p(t)$ is evaporation from the RP at time t [LT^{-1}]; $Q_o(t)$ is outflow rate of surplus runoff from the RP at time t [L^{3T-1}]; $H(t)$ is water depth at time t [L]; $H(t - \Delta t)$ is the water depth at the time, $t - \Delta t$ [L]; Δt is the time interval [T]; L_f is the length of the advancement of wetting front at time t [L]; $L_f(t - \Delta t)$ is the length of the advancement of wetting front at time $t - \Delta t$ [L]; K_s is the saturated hydraulic conductivity [LT^{-1}]; ψ_f is the suction head at the wetting front [L]; A_w is the area of the pond's catchment [L^2]; A_t is the top surface area of the RP [L^2]; $\bar{A}_{ws}(t)$ is the average water storage surface area between time $t - \Delta t$ and t [L^2]; and $\bar{A}_{rs}(t)$ is the average wetted planner area for recharge at time t [L^2].

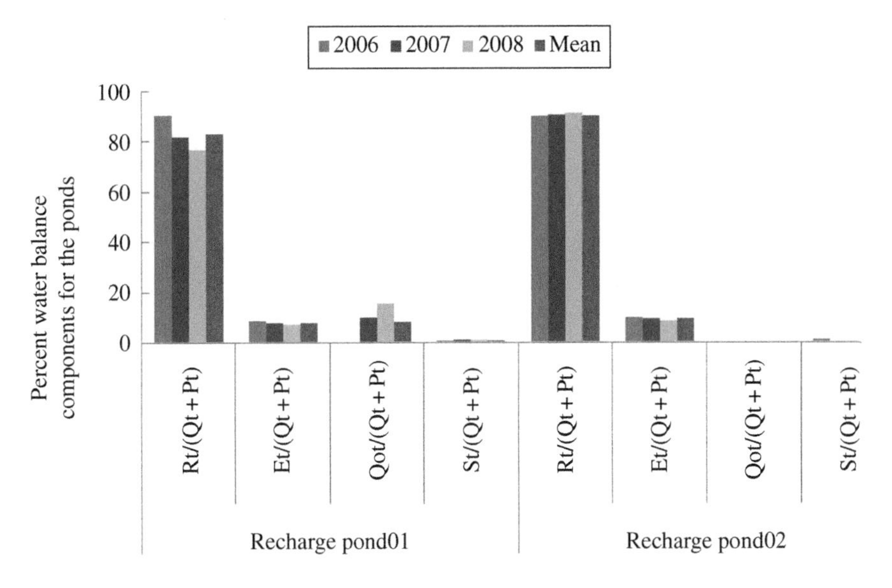

Figure 5.4 Partition factors of the water balance components (percent) for the ponds during the simulation period (2006–2008). Adopted and redrawn; Ali (unpublished PhD thesis, 2009).

Application of developed IPGRS model in Badakheda watershed, Rajasthan, India revealed that on an average 83.0–90.2% of accumulated surface runoff in the ponds contributed to potential GR into aquifer underneath ponds. Evaporation losses varied from 7.7 to 9.2% of stored runoff. Excess flows from the ponds and stored runoffs in ponds at the end of simulation periods ranged from 0 to 8.3 and 0.6 to 0.8%, respectively (Figure 5.4). Sensitivity analysis of potential recharge using the developed model suggested that an increase of inflows, which includes runoff from the pond's catchment and rainfall over the pond resulted in an increase in the annual volume of recharge and other associated variables (i.e., evaporation, outflow, hydroperiod, and wetting front length) and vice versa. Sensitivity analysis of RP indicated that an increase in the potential recharge, hydro-period, and length of the advancement of wetting front is lower than that of the evaporation and outflow, which ranged from 5 to 9, 1 to 2, and 4 to 7%, respectively, as the inflow increased from 12.5 to 75% to its original value. The increase in evaporation and outflow ranged from 28 to 29 and 50 to 474%, respectively. Conversely, reduction in the recharge, evaporation, hydro-period, and length of the advancement of the wetting front due to decreased inflow was relatively higher, which ranged from −4 to −75, −3 to −51, −4 to −51, and −1 to −52%, respectively (Figure 5.5). The increase in evaporation due to an increase in temperature had a small reduction in recharge (−1 to −4%), hydro-period (−1 to −4%), and length of the advancement of the wetting front (0 to −4%). The effect was slightly more pronounced on the outflow, which reduced from −3 to −13% as evaporation increased from 12.5 to 75%. The increase in the recharge, outflow, hydro-period, and length of the advancement of the wetting front due to a decrease in evaporation was found to be almost of the same magnitude as in the case of increased evaporation (Figure 5.6).

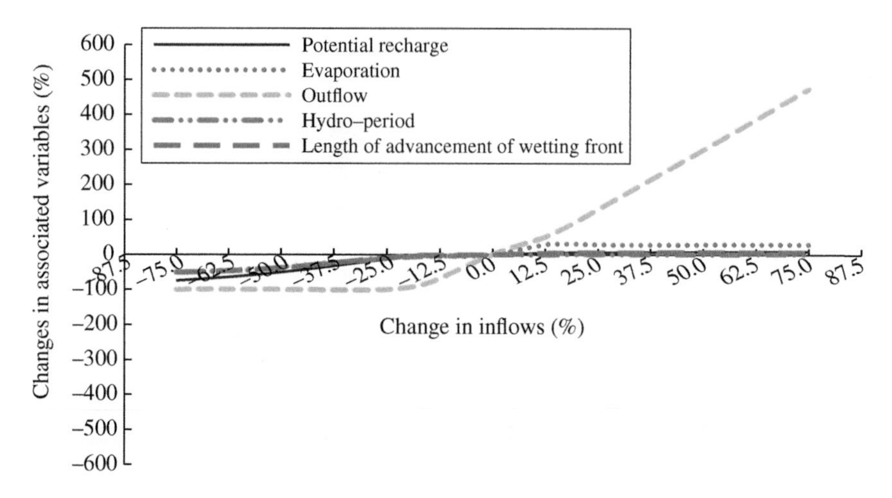

Figure 5.5 Effect of inflow on potential recharge, evaporation, outflow, hydroperiod, and length of the advancement of the wetting front in recharge pond01.

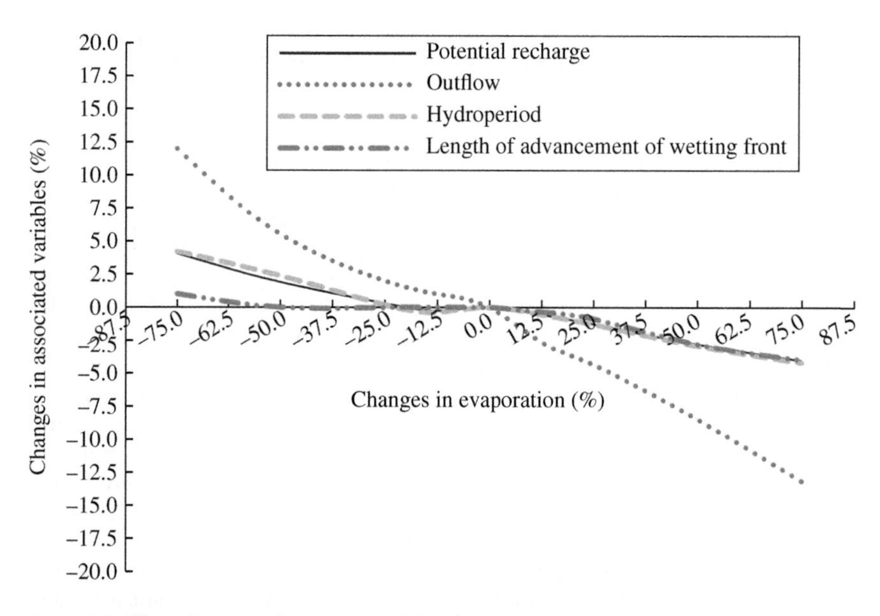

Figure 5.6 Effect of evaporation on potential recharge, evaporation, outflow, hydroperiod, and length of the advancement of wetting front in recharge pond01.

The $R_t/(Q_t + P_t)$ equation is the ratio of the total volume of water recharged into the aquifer, R_t, to the total volume of inflows, which is the sum of the volume of runoff into the pond and the volume of rainfall directly over the pond, $(Q_t + P_t)$; $E_t/(Q_t + P_t)$ is the ratio of the total volume of water loss by evaporation, E_t, to the total volume of inflows; $Q_{ot}/(Q_t + P_t)$ is the ratio of the total volume of outflows from the pond, Q_{ot}, to the total volume of inflows; and $S_t/Q_t + P_t$) is the ratio of the total volume of water remained as storage in the pond at the end of the simulation period, S_t, to the total volume of inflows.

Surface Water Evaporation

Evaporation is the largest component of water loss from the water storage system in an agro-ecosystem and has no beneficial use. It is one of the most complex components to model or measure in an agro-ecosystem being a function of solar radiation, temperature, wind speed, vapor pressure deficit, atmospheric pressure, and the surrounding environment. Numerous models exist for estimation of evaporation, which includes empirical (Singh & Xu, 1997), water budget (Abtew, 2001), energy budget (Priestley & Taylor, 1972; Sacks et al., 1994), mass transfer (MT) (Harbeck, 1962), and combination of energy and mass (Penman, 1948). While evaluating the performances of Bowen ratio energy balance (BREB), MT, Priestley–Taylor (PT), and pan evaporation (PE) models based on field data, the BREB model has been identified as the most effective and the most reliable evaporation estimation model for an agro-ecosystem in the semi-arid region of India when the Bowen ratio (β) is known a priori, and with limited data requirement (Ali et al., 2008). The identified BREB model is simple in use, is less data-driven, and can successfully be extended for free water surface evaporation estimates in the agro-ecosystem of the semi-arid regions of India. The simulated surface water evaporation from a pond in the agro-ecosystem by the BREB model ranged from 0.67 to 6.56, 1.61 to 4.94, and 2.61 to 3.07 mm d^{-1} for the daily, monthly, and seasonally (water available from first fill to its empty), respectively. In another study, Ali et al. (2016b) developed a suitable air–water temperature relationship to study the effects of increased water temperature on the small pond's attributes for a small aquatic pond in a semiarid region of India. They reported a 1.3–3.7 °C increase in pond water temperature with increases in air temperature from 1.5 to 4.3 °C by the end of 2080. This increase in water temperature resulted in an increase in the water evaporation rate by 8.3–30.3% and a decrease in the hydro-period and saturated dissolved oxygen by 3–26 d and 2.2–6.5%, respectively. This relatively simple modeling tool could be easily used by water resource managers for understanding and predicting changes in water temperature under climate change scenarios and preparing suitable management plans.

Potential Evapotranspiration

Evapotranspiration demand of an agro-ecosystem due to global warming will have a deep impact on agro-ecosystem provisioning services and irrigation water demand. The agro-ecosystem potential evapotranspiration depends mainly on air temperature, net solar radiation, relative humidity, and wind speed, as well as crop canopy characteristics (e.g., canopy height, leaf area index, and stomatal conductance) (Irmak et al., 2006; Martin et al., 1989). Due to global warming, increasing atmospheric CO_2 concentrations will have physiological effects on some crops and plants of agro-ecosystems, such as an increase in photosynthetic rate, leaf area, biomass, and yield, in addition to the reduction in stomatal conductance and transpiration per unit of leaf area (Allen, 1990; Kimball, 2007). Several studies have reported that effect of higher temperature on evapotranspiration could be either moderated or exacerbated by changes in the other climatic elements (radiation, vapor pressure deficit, humidity, and wind) and in plant factors (leaf area index, stomatal resistance) (Hsiao et al., 2019; Islam et al., 2012a). The CO_2–induced effects on potential evapotranspiration were found to be greater than those induced by the change in temperature for irrigated potato (*Solanum tuberosum* L.) agriculture in the San Luis Valley of south-central Colorado (Ramirez & Finnerty, 1996). Kirschbaum and McMillan (2018) also reported reductions in daily transpiration rates with higher atmospheric CO_2 concentrations.

There are several methods available for estimating potential or reference evapotranspiration. The FAO-56 Penman–Monteith (PM) is one of the most widely used methods for estimating reference evapotranspiration (ET_0) as the International Commission on Irrigation and Drainage (ICID) and the Food and Agriculture Organization (FAO) have recommended this method as the standard method of ET_0 estimation (Allen et al., 1998). This model explicitly incorporates both energy and biophysical parameters and is preferred for climate change impact studies, as it includes the effects of changes in the greater number of atmospheric variables (Kingston et al., 2009). Further, the effect of elevated CO_2 levels on evapotranspiration can also be accounted for by modifying the canopy resistance term in the PM model (Islam et al., 2012a; Wu et al., 2011). Priya et al. (2014, 2015) used FAO-56 PM to study the effect of climate change on evapotranspiration demand in Varanasi (India) and Akola (India). They reported a 6% decrease in annual ET_0 demand of Varanasi, India with doubling (660 mg kg^{-1}) of CO_2 concentration and temperature remaining constant. The effect of about 1.0 °C rise in temperature increased annual ET_0 by 2.3%, that was offset by an increase in CO_2 levels up to 495 mg kg^{-1}, and 2.5 °C rise in temperature increased ET_0 by 2.4%, which was offset by an increase in CO_2 levels up to 660 mg kg^{-1} (Priya et al., 2014). For Akola, there was about a 10% decrease in annual ET_0 demand with doubling (660 mg kg^{-1}) of CO_2 concentration with the temperature remaining constant. The effect of about 1.65 °C rise in temperature on annual ET_0 in Akola was offset by the increase in CO_2 levels up to 495 mg kg^{-1}, whereas the effect of 4 °C rise in temperature was offset by the increase in CO_2 concentrations up to 660 mg kg^{-1} (Priya et al., 2015). This shows that the CO_2 effect depends upon the type of crops in the area (e.g., wheat [*Triticum aestivum* L.] vs. maize [*Zea mays* L.]). As the effect of rising temperature due to global warming is moderated by the increasing CO_2 concentrations, it is essential to consider the effect of CO_2 concentration along with other climatic variables while estimating expected changes in irrigation water requirement due to global warming and planning for the development of future water resources and irrigation water management.

Surface Runoff

Runoff from agro-ecosystems causes water insecurity to surface soil and groundwater resources. Quantification of runoff in agro-ecosystems is a complex problem and is affected by a number of agro-ecosystem characteristics such as land use, land covers, and cover conditions (Ali et al., 2010), topography (Tejpal, 2013), morphology (Ali & Singh, 2001), antecedent soil moisture condition (Brocca et al., 2008; Kim et al., 2005; Wang et al., 2007), density of protective and conservation measures (Ali et al., 2017), and rainfall amount, intensity, and duration (Wang et al., 2007). The understanding of runoff processes and their quantification is essential for water conservation and further emphasized by the need to develop adaptations for climate change and its variability and desire to develop climate-resilient agro-ecosystem practices. Many runoff predicting models based on physical, statistical, conceptual, and/or combination approaches have been developed. These models have been used in the past for better understanding and simulating the extremely complex, nonlinear dynamic rainfall–runoff process according to need and availability of data (Haan et al., 1995; Singh & Woolhiser, 2002; Wang et al., 2007). The commonly used methods include Soil Conservation Service Curve Number (SCS-CN; USDA-SCS, 1993), Hydrologic Engineering Center–Hydrologic Modeling System (HEC-HMS; Feldman, 1981), Chemicals, Runoff, and Erosion from Agricultural Management Systems (CREAMS; Knisel, 1980), Kinematic Runoff and Erosion Model (KINEROS; Woolhiser et al., 1990), soil–vegetation–atmosphere transfer (SVAT; Arora, 2002), Soil Water Assessment Tool (SWAT; Arnold et al., 1998), and Precipitation Runoff Modeling Systems (PRMS; Leavesley et al., 2002).

Ali et al. (2010) developed a simple rainfall–runoff model based on the normalized antecedent precipitation index (NAPI) using the water balance concept in the agro-ecosystem. The model is defined as:

$$Q = \frac{P\left(-bP + c\,\text{NAPI} + a\right)}{\left[\left(-bP + c\,\text{NAPI} + a\right) - 1\right]} \text{ valid for } P > 0 \tag{5.6}$$

where Q is the runoff for corresponding rainfall P [L]; NAPI is [–]; a [dimensionless], b [dimension of L^{-1}], and c [dimensionless] are coefficients specific to a watershed.

The NAPI is defined as (Ali et al., 2010):

$$\text{NAPI} = \frac{\sum_{t=0}^{-i} P_t k^{-t}}{\bar{P} \sum_{t=-1}^{-i} k^{-t}} \tag{5.7}$$

where i is the number of antecedent days; k is the decay constant [–]; P_t is the rainfall during t day [L]; $\bar{P}$ is the average rainfall for antecedent days [L]; and i is usually taken every 5, 7, or 14 d and the value of k ranges between .80 and .98.

The derived model is simple, easy to use, and very less data-driven. Only one input variable, rainfall, and rainfall-derived NAPI is needed, if model parameters, a, b, and c are a priori known. The developed model is a very useful tool for predicting runoff yields in the agro-ecosystem and has wide applicability. It can also be used for runoff estimation from gauged and ungauged watersheds with limited data of rainfall only. The runoff simulating potential of the derived model has been evaluated with the SCS-CN model, and results indicated that the derived model compared well with the SCS-CN model, which is a relatively more data-demanding model (Figure 5.7). The derived model has also been applied in three small agricultural watersheds located in the ravine area in a semi-arid agro-ecosystem of Rajasthan, India (Figure 5.8). The simulated runoff for the cropped and ravine area of the agro-ecosystem ranged from 10 to 20% of the rainfall (Ali et al., 2010).

There is growing interest in the application of different hydrological models in River basins of India to assess climate change impact on hydrology and water resources, mainly using hydrological models coupled with General Circulation Model (GCM) projections (Islam et al., 2014). Passcheir (1996) compared 5-event

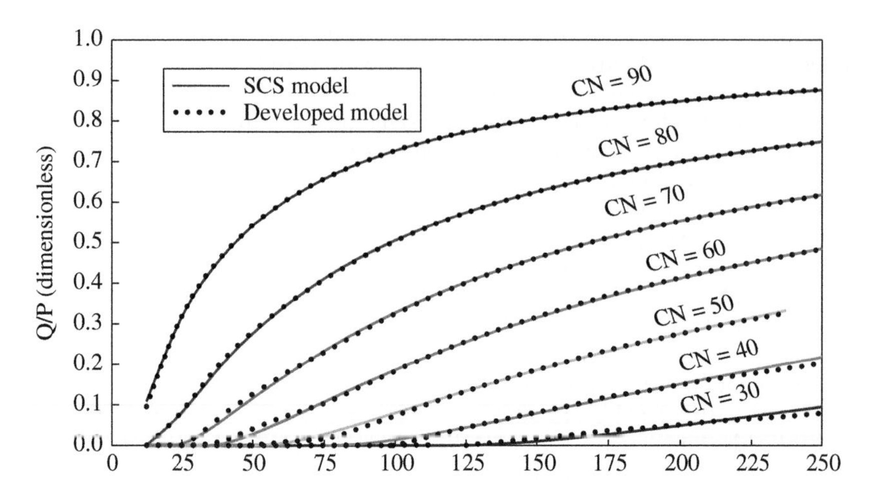

Figure 5.7 Comparison of the generated *Q*/*P* (runoff/rainfall) profiles by the SCS-CN model for P and different CN values with the corresponding profiles of the developed model. SCS, Soil Conservation Service; CN, curve number.

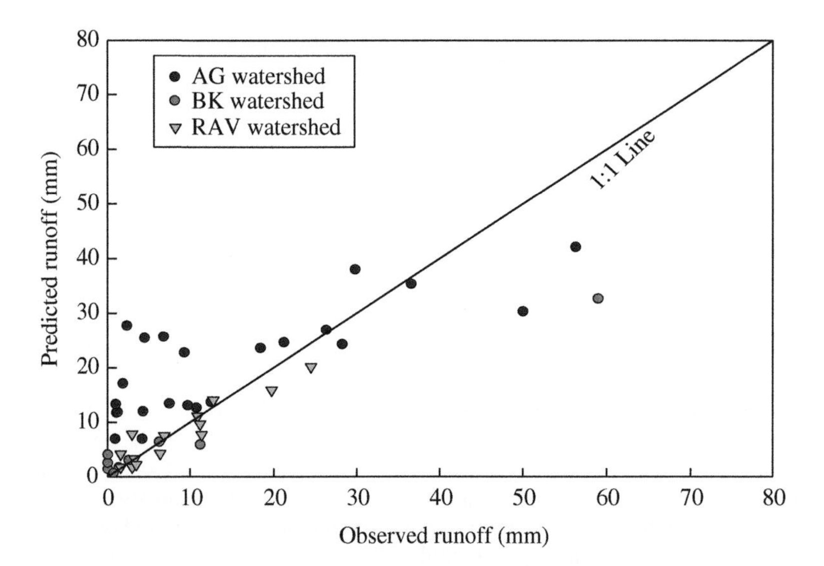

Figure 5.8 Comparison of observed and predicted runoff by the developed model in three small agro-ecosystem watersheds in the semi-arid region of India. AG, agricultural; BK, Badakheda; RAV, ravine. Adopted from Ali et al. (2010).

(single runoff event) models and 10 continuous hydrological models for rainfall–runoff modeling of the Rhine and Meuse basin for land use impact modeling, climate change impact modeling, real-time flood forecasting, and physically based flood frequency analysis. Four continuous models, namely, PRMS, SACRAMENTO, HBV, and SWMM, and one event model (HEC-1) were evaluated as the best ones.

The HEC-1 and HBV models were found to be the most appropriate for flood frequency analysis, the HBV and SLURP models for climate change impacts on peak discharges, and the PRMS and SACRAMENTO model for assessment of climate change impact on discharge regimes.

The US Geological Survey's PRMS (Leavesley et al., 2002) has been widely used to study the effect of land use and climate change scenarios on streamflow (Islam et al., 2012b; Qi et al., 2009). Islam et al. (2012b) applied PRMS to simulate climate change impact on streamflow in the Brahmani River basin of India and reported a 62% increase in annual streamflow under the combined effect of 4 °C temperature rise, and 30% rainfall increase in the Brahmani River basin. Recently, Vandana et al. (2018) applied PRMS to study the impact of climate change on streamflow in the Brahmani River basin using multi-model ensemble climate change scenarios. The multi-model ensemble climate change scenarios were generated from 16 Coupled Model Inter-comparison Project phase 3 (CMIP3) GCMs projections under three different emission scenarios of A2 (high emission), A1B (medium emission), and B1 (low emission scenarios). They reported changes in annual streamflow in the range of 2.4–4.7 and 7.3–12.6% during the 2050s and 2080s, respectively (Figure 5.9).

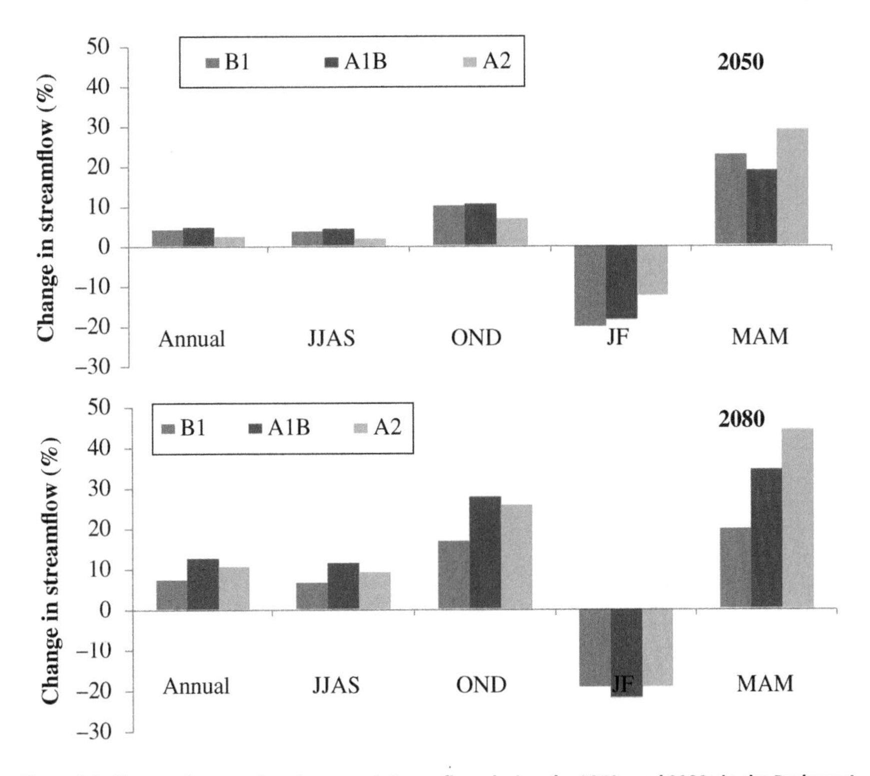

Figure 5.9 Changes in annual and seasonal streamflow during the 2050s and 2080s in the Brahmani River basin of India. JJAS, June–July–August–September; OND, October–November–December; JF, January–February; MAM, March–April–May; B1, A1B, and A2 are low to high greenhouse gas (GHG) emission scenarios.

Further, there is an increase in streamflow during monsoon (JJAS), post-monsoon (OND), and pre-monsoon (MAM) season also. Changes in streamflow during monsoon season varied in the range of 1.8–4.2 and 6.6–11.4% during the 2050s and 2080s, whereas during the post-monsoon period, it varied in the range of 6.9–10.5 and 16.8–27.8% during the future periods of 2050s and 2080s, respectively. During the winter season (JF), there is a decrease in streamflow during all three future periods and it varied in the range of 12.3–20.2%, and 19.1–21.9% during the 2050s and 2080s, respectively. As there is temporal variation in the streamflow in the basin, there is a need for developing different irrigation water management adaptation strategies for crop planning. Their simulation results also showed an increase in the magnitude of flood flows (Q5) and a decrease in the magnitude of low flows (Q95) under all scenarios during future periods (Figure 5.10). The increase in high flows (Q5) ranged from 1.7–4.5 and 6.8–12.1% during the 2050s and 2080s, respectively. This increase in high flow was greater under A1B emission scenarios. The increase in frequencies of high flow events varied in the range of 4.9–11.2 and 17.6–33.9% during the 2050s and 2080s, respectively. The low flow (Q95) indices indicated a decrease in the low flows in the basin during future periods under all the emission

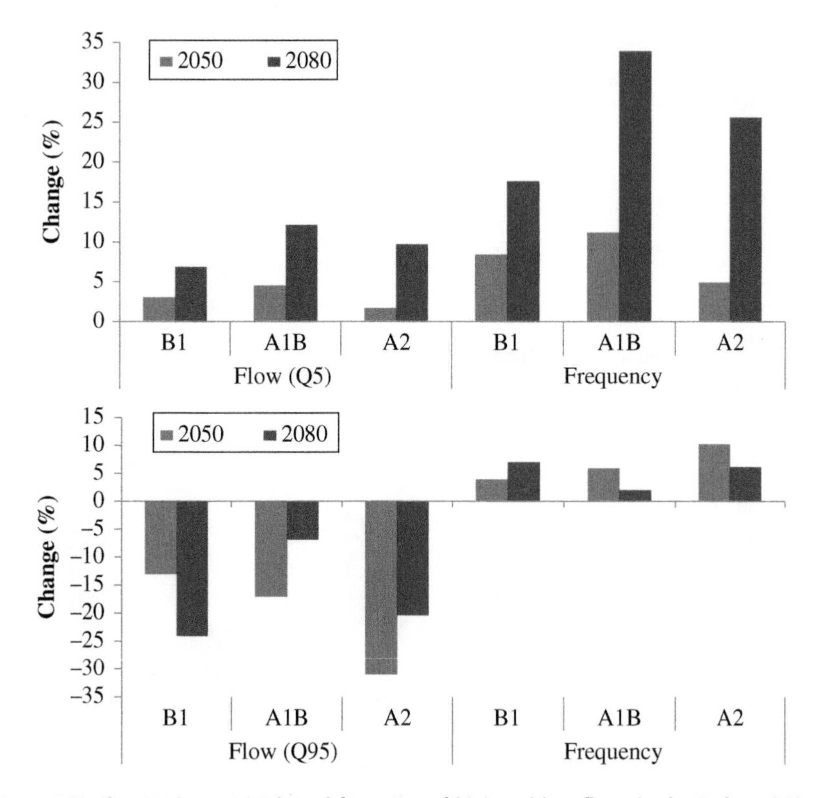

Figure 5.10 Changes in magnitude and frequency of high and low flows in the Brahmani River basin. Q5, high flows; Q95, low flows; B1, A1B, and A2 are low to high greenhouse gas (GHG) emission scenarios.

scenarios, and this decrease in low flows varied in the range of 13.1–31.0 and 6.9–24.1% during the 2050s and 2080s, respectively. The frequency of low-flow events increased in the range of 3.9–10.2 and 2.0–6.1% during the 2050s and 2080s, respectively (Figure 5.10). This decrease in low flow is likely to impact the environmental flow requirement of the Brahmani River. Further, an increase in magnitude and frequency of high (flood) flows is likely to have severe impacts, particularly in the coastal regions of the basin.

The SWAT is one of the most widely used models for simulating basin hydrology and assessing the effect of land use and climate change at basin scale (Arnold et al., 1998). The major model components include hydrology, erosion/sedimentation, crop growth, nutrient, pesticide, agricultural management, channel and pond/reservoir routing, and weather generation (Arnold et al., 1998) and has been widely used for long-term simulations in agricultural watersheds. The SWAT model has also been used for quantifying ecosystem services and estimating the consequences of management impacts (Francesconi et al., 2016; Vigerstol & Aukema, 2011). Francesconi et al. (2016) provided an excellent overview of the application of SWAT to quantify ecosystem services, model's capability in examining various types of ecosystem services, and also described the approaches used by various researchers to model different ecosystem services. The SWAT model can be used to estimate soil erosion from the watershed and sediment deposition, water quality, nutrient, and pesticide as regulating ecosystem services, to estimate evapotranspiration to maintain soil moisture condition as an ecosystem service to improve water conservation and so forth. Francesconi et al. (2016) also reported that SWAT may preferably be used to simulate provisioning and regulating services, and proxy variables can be used to estimate associated supporting and cultural services.

The SWAT model has been applied by several researchers to investigate the climate change impacts on hydrology and water resources availability in Indian River basins (e.g., Abeysingha et al., 2018; Gosain et al., 2011; Mishra & Lilhare, 2016; Narsimlu et al., 2013) and simulating crop production (Bhuvaneswari et al., 2013; Lakshmanan et al., 2011) using different GCM projected climate change scenarios. Lakshmanan et al. (2011) modeled the hydrology and rice (*Oryza sativa* L.) yield of the Bhavani basin in Tamil Nadu, India, and showed that the SWAT model can be employed under different climate change and management scenarios for developing adaptation strategies to sustain rice production under changing climate scenarios. Abeysingha et al. (2016) applied SWAT for assessing future impacts of climate change on irrigated rice and wheat production and their evapotranspiration and irrigation requirements in the Gomti River basin (India). In a recent study, Abeysingha et al. (2018) applied the SWAT model for assessing climate change impact on water availability in the Gomati River, an important tributary of the Ganges River (India). They reported an increase in annual streamflow in the range of 1.1–6.9 and 2.1–14.7% during the 2050s and 2080s, respectively, under different representative concentration pathways (RCPs) (Figure 5.11). Seasonal analysis showed an increase in streamflow in the monsoon (June–September) and in the post-monsoon (October–December) seasons and decrease in streamflow in the winter (January–February) and summer (March–May) seasons in the Gomti River basin. This increase in annual streamflow is mainly due to

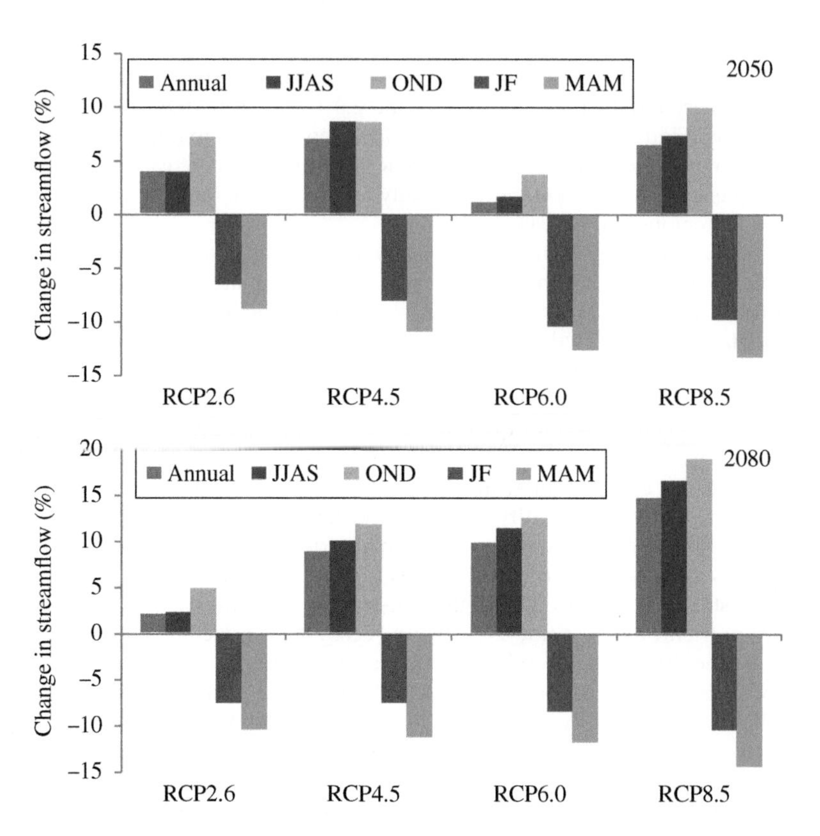

Figure 5.11 Changes in annual and seasonal streamflow during the 2050s and 2080s in the Gomti River basin. RCPs, representative concentration pathways; JJAS, June–July–August–September; OND, October–November–December; JF, January–February; MAM, March–April–May.

an increase in rainfall in the basin. The annual rainfall is projected to increase by 4.9–9.7 and 5.3–18.0% during the 2050s and 2080s, respectively, under different RCPs, and the basin is likely to experience a decrease in rainfall during the winter season. Surface runoff from the Ganges River basin is more sensitive to the changes in precipitation than that of temperature. Increased precipitation along with warming climate will largely lead to an increase in surface runoff in all the sub-basins in the Ganges River basin (Mishra & Lilhare, 2016).

Simulation results also showed an increase in high flows (Q5) and its frequency in the basin during the future periods of the 2050s and 2080s under all the four RCPs. The increase in high flows, as compared to baseline, varied in the ranges of 6.0–14.4 and 6.8–27.3% during the 2050s and 2080s, respectively. Further, the frequency of occurrence of high (flood) flows was also found to increase, and it varied in the range of 19.5–47.5 and 24.3–87.4% during the 2050s and 2080s, respectively. Changes in low flow (Q95) varied in the range of −3.2–0.0%, and mostly remaining within a ±5% range (Figure 5.12). There was also an increase in the frequency of low flow days and varied in the range of

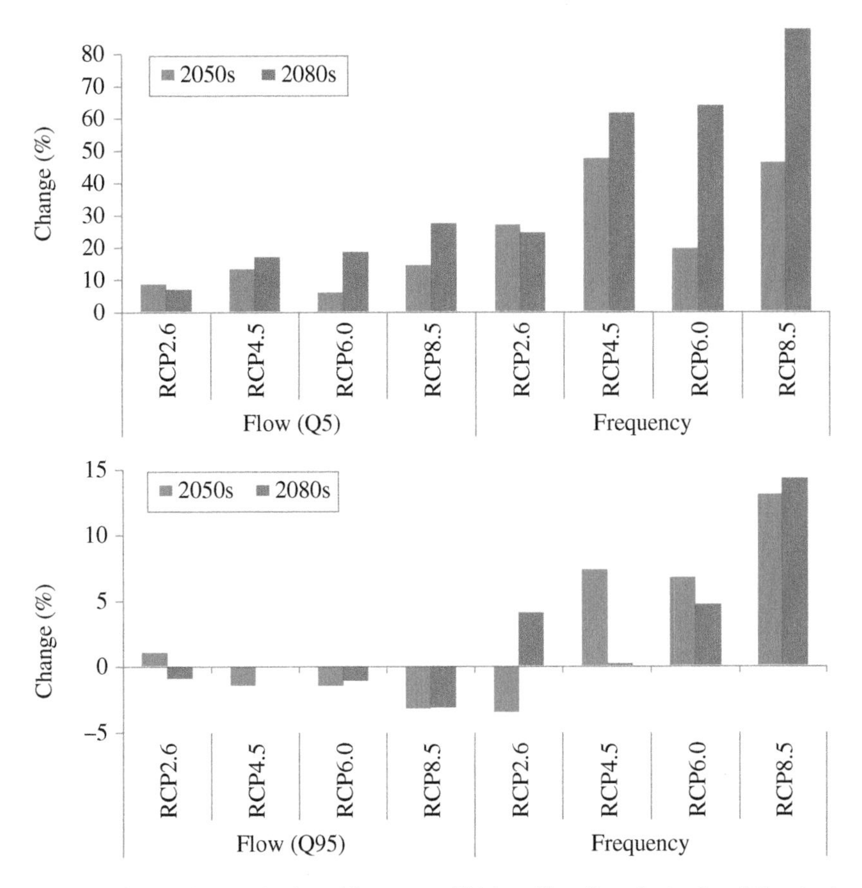

Figure 5.12 Changes in magnitude and frequency of high and low flows in the Gomti River basin. RCPs, representative concentration pathways.

−3.5–13 and 0.2–14% during the 2050s and 2080s, respectively. Thus, the basin faces the dual problem of excess water during the monsoon season and water scarcity conditions during the nonmonsoon season. The increase in the magnitude of high (flood) flow and the number of flood (high) flow days in the basin under future climate change scenarios may likely damage crops, livestock, and other properties, indicating the need for suitable adaptation measures, both engineering and agronomic, to avoid losses and damage to life and crops. Changes in low flow in the Gomti River basin may not be of immediate concern; however, preparing a long-term adaptation plan with due attention to maintain environmental flows and ecosystems services is necessary as several tributaries of the Gomti River basin are dry during the non-monsoon months (Dutta et al., 2015). As an adaptation option, water harvesting and storing excess water in ponds and reservoirs during the wet season will not only mitigate water scarcity problems during the dry season but will also help to reduce the flood risk during monsoon season and improve GR (Sikka et al., 2018).

Surface Water Storage

Natural or anthropogenic surface water bodies play a pivotal role in maintaining the ecological and hydrological balance of an agro-ecosystem, especially by increasing water availability for extended periods. The time-variant water availability in surface water bodies plays a vital role in comprehensive and coordinated planning for the utilization of water resources in the agro-ecosystem. Several models have been developed in the past to simulate water depth or volume of water in surface water bodies such as dynamic linear predictor models (Kakahaji et al., 2013), nonlinear intelligence models (Altunkaynak, 2007; Kakahaji et al., 2013), and modeling code such as SPAW (soil–plant–air–water; Saxton & Willey, 2006).

Ali et al. (2015) developed a holistic water depth simulation (HWDS) model by incorporating the models for the rainfall–runoff, evaporation, and length of the advancement of wetting front in the water balance equation of the surface water body. As in the IPGRS model for simulating GR, the derived HWDS model also includes time-varying rainfall, runoff, surface water evaporation, outflow, and length of the advancement of wetting front; and saturated hydraulic conductivity and fillable porosity of the surface water body's bed material as their parameters. The HWDS is defined as:

$$H(t)=H(t-\Delta t)\frac{A_{\mathrm{ws}}(t-\Delta t)}{A_{\mathrm{ws}}(t)}+\frac{\Delta t}{A_{\mathrm{ws}}(t_i)}\left[Q(t)A_{\mathrm{w}}+P(t)A_{\mathrm{s}}-E(t)\bar{A}_{\mathrm{ws}}(t)-Q_0(t)\right]$$

$$-\frac{K_{\mathrm{s}}\,\bar{A}_{\mathrm{rs}}(t)\Delta t}{A_{\mathrm{ws}}(t)}\left[1+\frac{H(t)+\psi_{\mathrm{f}}}{L_{\mathrm{f}}(t-\Delta t)+\{H[t-\Delta t]+\psi_{\mathrm{f}}\}\left\{\sqrt{\frac{F_1K_{\mathrm{s}}}{\eta[H(t-\Delta t)+\psi_{\mathrm{f}}]}t_i+F_2}-\sqrt{\frac{F_1K_{\mathrm{s}}}{\eta[H(t-\Delta t)+\psi_{\mathrm{f}}]}(t-\Delta t)+F_2}\right\}}\right] \tag{5.8}$$

and

$$V(t)=H(t)\bar{A}_{\mathrm{ws}}(t) \tag{5.9}$$

where $V(t)$ is the volumetric water availability in a surface water body at time t_{i} [L^3]; $\bar{A}_{\mathrm{wa}}$ is the average water storage area at time t_{i} [L^2] $= 0.5[A_{\mathrm{ws}}(t) + A_{\mathrm{b}}]$; and A_{b} is the water body's bottom surface area [L^2], and other terms are defined earlier.

The HWDS model is process-based, holistic in nature, and easy to use compared with the other existing models. It can simulate time-variant water depth and volumetric water availability with reasonable accuracy from the water harvesting structures under variable water depth. The model has been successfully applied in the agro-ecosystems of India. The model has wide field applicability and can

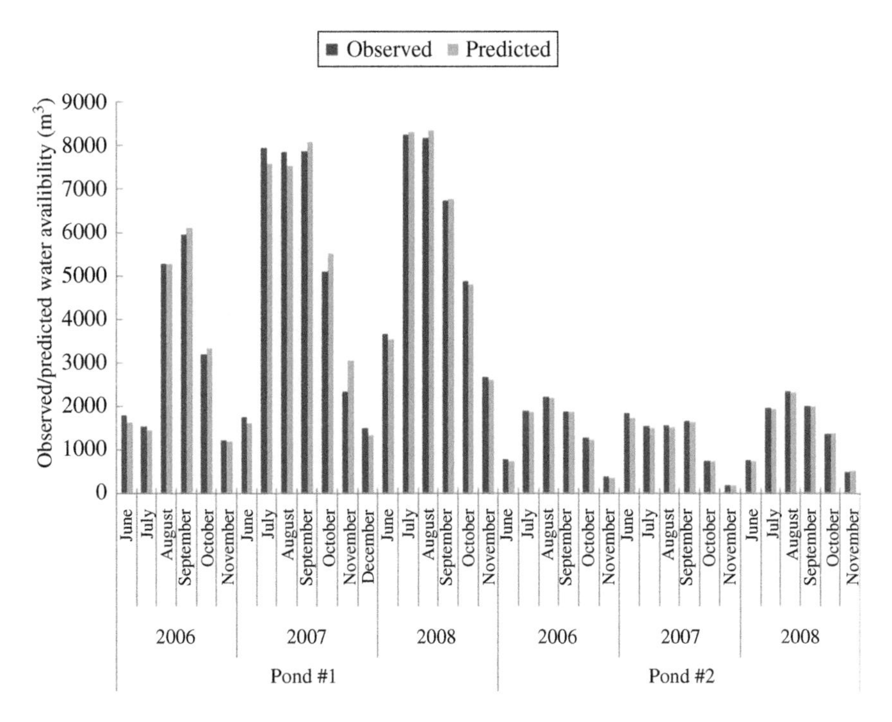

Figure 5.13 Comparison of time-variant observed and predicted water availability by the developed model in pond01and pond02 an agro-ecosystem watershed in the semi-arid region of India. Adopted and modified; Ali (2016).

successfully be extended for estimation of water depth and water availability in the water body of any shape and size, located in any geographical region of India and elsewhere. The simulated time-variant water availability by the derived model compared well with ponds data of the Badakheda watershed agro-ecosystem in Rajasthan of India (Figure 5.13) (Ali, 2016). The simulated time-variant water availability in the pond01 and pond02 was recorded higher in July, August, and September and almost empty in November, December, or both.

Soil Erosion

Soil erosion due to water is one of the major threats affecting society and the economy in different parts of the world (Mondal et al., 2014b; Pimentel, 2006). Soil erosion causes on-site problems with removal of the productive soil top layer (Pimental & Sparks, 2000), significant loss of agricultural land, deterioration of soil physical, chemical, and biological properties (Lal et al., 2000), loss of nutrients (Lal, 2003), loss of soil organic matter (Longbottom et al., 2014), and reduction of agricultural productivity as soil loss diminishes soil water storage capacity, nutrient, impacting crop growth, and flood risk (Lal, 1999). Soil erosion also brings about offsite damages, such as fluvial sediment deposition, reservoir sedimentation, channel/stream silting (Mullan, 2013), and surface water quality by agro-chemicals and colloid facilitated transport (Cochrane & Flanagan, 1999). Various factors that affect the rate of soil erosion involve anthropogenic factors such as an intensification of agro-ecosystem, land-use changes, and human activities and climate change, particularly changes in rainfall (Vereecken et al., 2016). The impacts of change in rainfall on soil erosion are mainly caused by changes in rainfall amount (Nearing et al., 2004; Bangash et al., 2013), rainfall intensity (Tang et al., 2015; Zhang et al., 2012), and spatio-temporal distributional patterns of rainfall (Maeda et al., 2010). Many researchers assessed the soil erosion problems (Ali & Sharda, 2005; Dabral et al., 2008; Pandey et al., 2009; Sharda & Ali, 2008) and impact of climate change on soil erosion (Khare et al., 2016; Mondal et al., 2014a, 2016) using the various soil erosion models.

It has been estimated using the conceptual model that about 5.11×10^9 Mg of soil is being detached annually in India due to different reasons (Sharda & Ojasvi, 2016), out of which 34.1% of the total eroded soil is deposited in the reservoirs, 43.0% is displaced within the river basins, and the remaining 22.9% is discharged outside the country mainly to oceans. By using a conceptual modeling framework, Sharda and Ojasvi (2016) estimated that the northern India river basins contribute about 81% of the total sediment yield from the landmass and 19% by the southern river basins. Soil erosion from agricultural land under different crop and cropping systems and soil type are reported to vary in the range of 2.5–64.5 Mg ha^{-1} yr^{-1} (Ali & Sharda, 2005; Dhruva Narayana & Babu, 1983; Sharda et al., 2010). Long-term average annual and annual soil erosion rates predicted by the universal soil loss equation (USLE) model for different cropping systems in the sub-humid climates of India ranged from 1.26 to 4.03 and 0.34 to 6.10 Mg ha^{-1} yr^{-1}, respectively, whereas for the semi-arid region respective soil erosion was of 0.06–0.41 and 0.01–0.96 Mg ha^{-1} yr^{-1} (Ali & Sharda, 2005). The mean of long-term average annual and annual erosion rate by the USLE model was comparable with observed values through field experiments (Figure 5.14). A severe rate

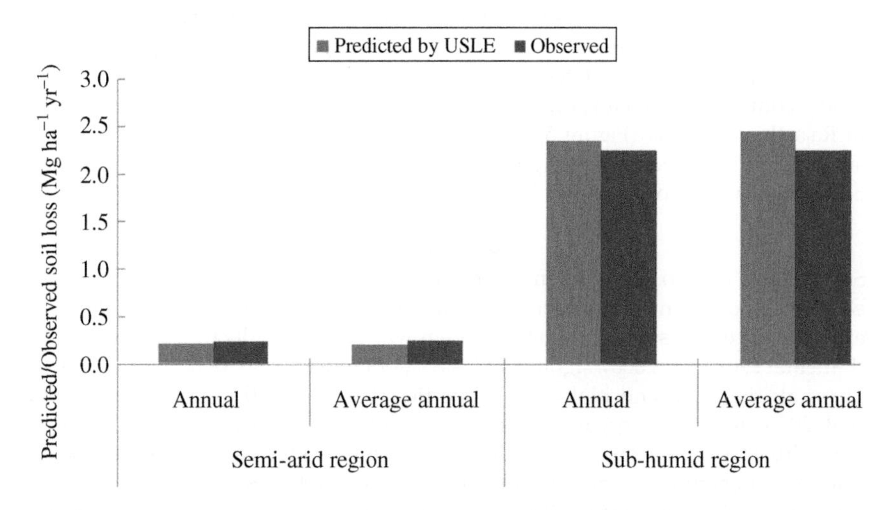

Figure 5.14 Comparison of mean long-term average annual and annual observed and predicted soil erosion by the USLE in the semi-arid and sub-humid region of India. Adopted and redrawn; Ali and Sharda (2005).

(57.1–75.7 Mg ha^{-1} yr^{-1}) of soil loss in the Dikrong River basin of Arunachal Pradesh was reported by Pandey et al. (2009) and Dabral et al. (2008) using the USLE and Morgan–Morgan–Finney (MMF) models, respectively. Similarly, Biswas and Pani (2015) applied the revised universal soil loss equation (RUSLE) model and reported moderate to high soil erosion (14–282 Mg ha^{-1} yr^{-1}) soil erosion over the Barakar River basin in Jharkhand State.

Ali et al. (2002) developed a simple soil erosion prediction model for estimating potential soil erosion from an agro-ecosystem. The derived model is defined as:

$$Y = a(RKC)^b \tag{5.10}$$

where Y is the annual soil loss (Mg ha^{-1}), R is the annual rainfall erosivity factor, K is the soil erodibility factor (Mg ha^{-1} unit of rainfall erosion index at maximum 30 min rainfall intensity, IE_{30}), C is the crop and cover management factor, and a and b are the model parameters specific to an agricultural field.

The derived model has reasonable accuracy, is simple in nature, easy to use, and very less data-driven. The developed model is a very useful tool with limited data sets for predicting potential soil erosion yields from the semi-arid agro-ecosystem of India and has wide applicability. The model has also been tested in the field for various crop and cropping systems in semi-arid regions of Rajasthan (India) and found to perform well (Figure 5.15). The predicted soil erosion by the derived model for the grasses (i.e., bluestem [*Dichanthium annulatam* Willemet], bermudagrass

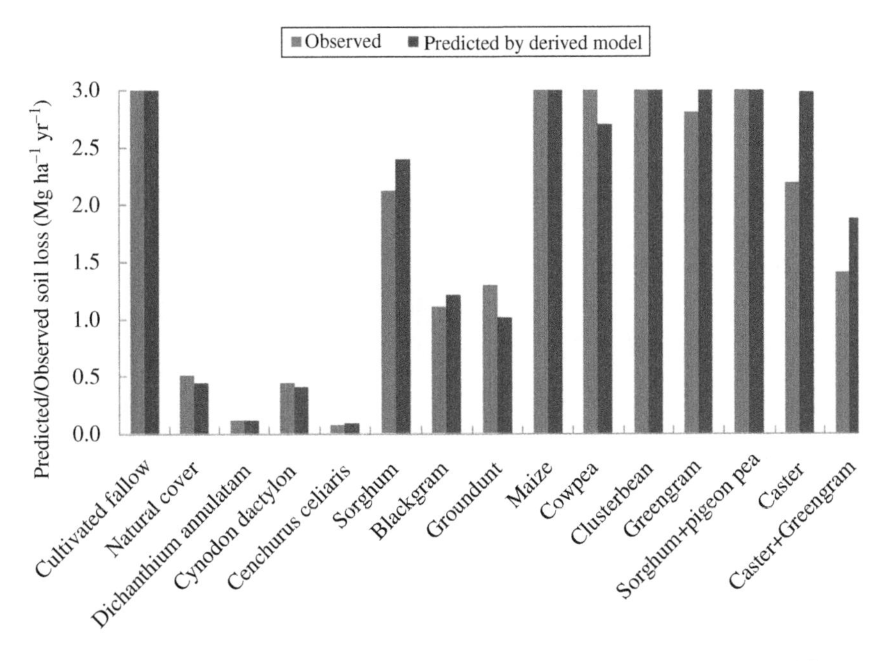

Figure 5.15 Comparison of observed and predicted soil erosion by the developed model for different crop and cropping system agro-ecosystem in the semi-arid region of India. Adopted and modified; Ali et al. (2002).

[*Cynodon dactylon* (L.) Pers.], and buffelgrass [*Cenchurus celiaris* L.]), close-growing crops (black gram [*Vigna mungo* (L.) Hepper], groundnut [*Vigna subterranea* (L.) Verdc.], cowpea [*Vigna unguiculata* (L.) Walp.], and green gram [*Vigna radiata* (L.) R. Wilczek]) ranged from 0.09 to 0.11 and 1.01 to 3.05 Mg ha^{-1} yr^{-1}, respectively. However, higher soil erosion was recorded for the broad spacing (maize, sorghum [*Sorghum bicolor* (L.) Moench], clusterbean [*Cyamopsis tetragonoloba* (L.) Taubert], and castor [*Ricinus communis* L.] and mixed (sorghum–pigeon pea and caster–green gram) cropping systems, which ranged from 2.39 to 4.22, and 1.88 to 3.83 Mg ha^{-1} yr^{-1}, respectively. The highest erosion of 4.46 Mg ha^{-1} yr^{-1} was simulated for the cultivated fallow, and 0.44 Mg ha^{-1} yr^{-1} for the natural cover.

In another study, Mondal et al. (2016) simulated climate change impact on soil erosion in the Narmada basin of India using the RUSLE model coupled with the HADCM3 model simulated climate change scenarios (Hadley GCM 3) under SRES A2 emission scenarios. Simulated soil erosion for different land uses, soils, and land slopes increased from 58 to 59, 52 to 63, and 53 to 59% during the 2020s, 2050s, and 2080s, respectively, over the base period (1961–2001) (Mondal et al., 2016). An increase in annual rainfall amount by 35.73% in the 2080s resulted in increased erosion by 58.37% and a soil organic C loss by 56.55% during the 2080s. The base period (1961–2001) soil erosion rates under different land use, soils, and land slopes ranged from 3.51 to 12.52, 4.96 to 30.10, and 6.05 to 50.01 Mg ha^{-1} yr^{-1}, respectively (Figure 5.16). They also reported that simulated sediment load decreases by 5.3% in the 2020s over the base period sediment load (22.25 Mg yr^{-1}); however, simulated

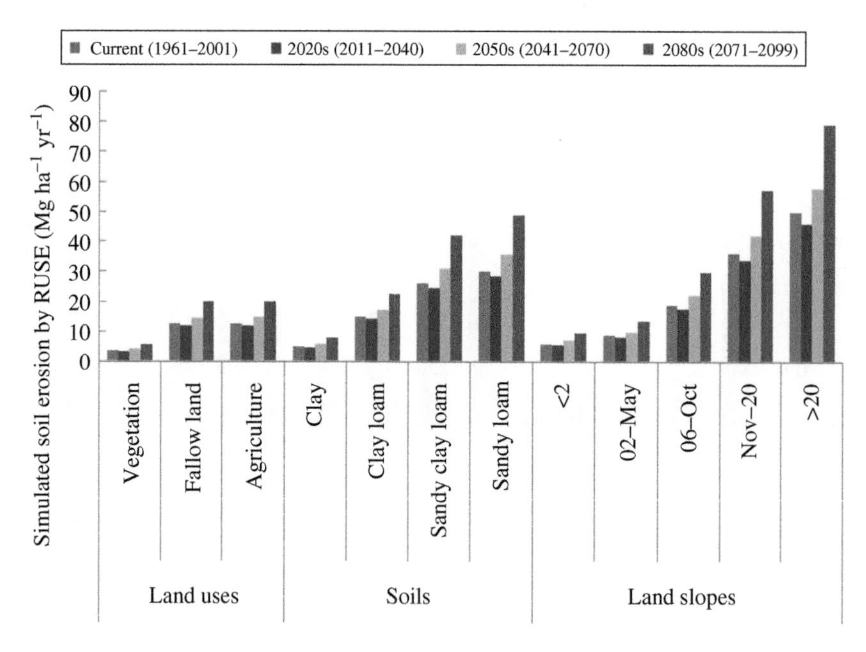

Figure 5.16 Simulated soil erosion rate by the RUSLE for different land uses, soils and land slopes in Narmada river basin during the base period (1961–2001), the 2020s (2011–2040), 2050s (2041–2070), and 2080s (2071–2099) under climate scenario A2. RUSLE, revised universal soil loss equation. Adopted and modified; Mondal et al. (2016).

sediment load increases about 18 and 58.4% in the 2050s and 2080s, respectively (Mondal et al., 2016).

In a similar study for Himalayan regions of India, Gupta and Kumar (2017) simulated soil erosion using the RUSLE model coupled with HadCM3 projected climate change scenarios under A2 and B2 emission scenarios. The average annual soil loss was projected to increase by 28.38, 25.64, and 20.33% during the 2020s, 2050s, and 2080s, respectively, under the A2 scenario, whereas under the B2 scenario, it was projected to increase by 27.06, 25.31, and 23.38% during the 2020s, 2050s, and 2080s, respectively, from the base period (1985–2013). Further, the highest soil erosion rate was projected in the uncultivable wastelands (197.0–202.0 Mg ha^{-1} yr^{-1}), followed by scrublands (66.7–70.4 Mg ha^{-1} yr^{-1}) during the 2020s (2011–2040) and 2050s (2041–2070) under A2 and B2 scenarios (Figure 5.17). The uncultivable wastelands are the lands that are not available for cultivation such as barren rocky lands and steeply sloping areas. However, the lowest erosion rate was in the croplands (25.2–26.6 Mg ha^{-1} yr^{-1}). Under the A2 emission scenario, soil erosion for cropland increased by 28.7, 25.8, and 20.6% during the 2020s, 2050s, and 2080s, respectively, as compared to the base period (1985–2013) with soil erosion rate of 20.90 Mg ha^{-1} yr^{-1}, but increased during the 2020s, 2050s, and 2080s under the B2 scenario of 27.3, 25.5, and 24.0%, respectively. Moderately high soil erosion from the croplands in the mid-Himalaya area during the base period was mainly due to intense rainfall events and higher land slope. Results indicated need for adoption of appropriate soil and water conservation measures (i.e., terracing, growing close spacing crops, and so forth) to bring the soil erosion rate within the soil loss tolerance limit (≤2.5 Mg ha^{-1} yr^{-1}).

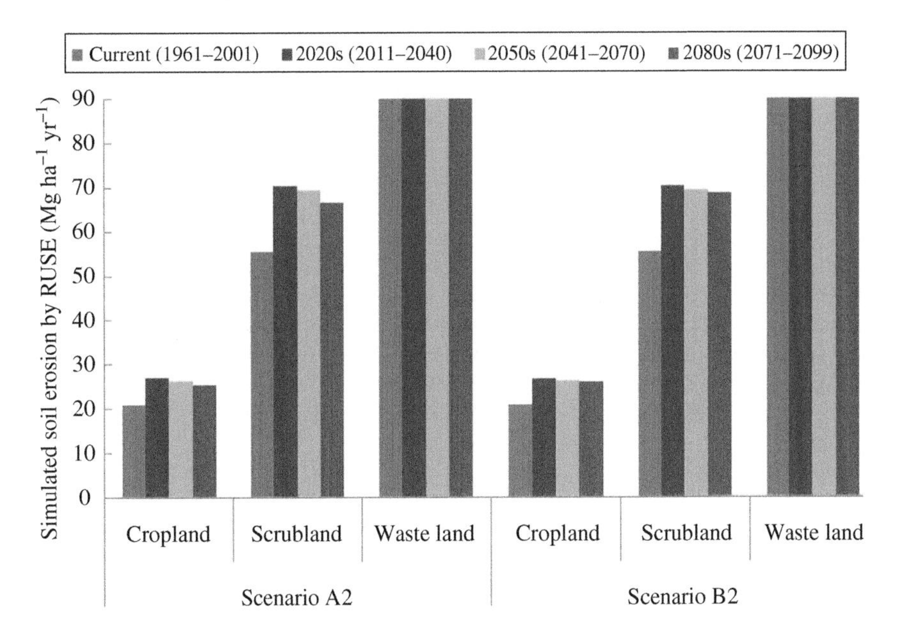

Figure 5.17 Simulated soil erosion rate by the RUSLE for the Himalayan regions of India during the base period (1961–2001), the 2020s (2011–2040) 2050s (20741–2070), and 2080s (2071–2099) under climate scenario A2 and B2. RUSLE, revised universal soil loss equation. Adopted and modified; Gupta and Kumar (2017).

Summary

The agro-ecosystem is one of the most important ecosystems on the earth, and it provides the four Fs (i.e., food, fiber, fodder, and fuel) to the more than 7.7 billion people in the world. Agro-ecosystems also produce a variety of water and climate-regulating services and cultural agro-ecosystem services. These agro-ecosystem services rely on supporting ecosystem services provided by natural ecosystems. However, the sustainability of agro-ecosystem services in India is threatened by climate change and variability, and offsets mostly negative climate change effects on the functional agro-ecosystem services. In order to understand and quantify the effects of climate change and variability on agro-ecosystem services, agro-ecosystem models have the potential to do it in a better way. This chapter presents some promising examples of applications of agro-ecosystem models in India to simulate some of the water-regulating agro-ecosystem services as infiltration, potential recharge, evaporation, evapotranspiration, surface runoff, water storage, and soil erosion, focused on climate change and variability. Simulation models have been successfully applied for these agro-ecosystem services in various agro-ecosystems of India.

Application of an infiltration model indicated that the time delays for wetting front to reach the depth to the water table (1–50 m) ranged from 1 h to several months in most of the soil textural classes. Application of the HYDRUS-1D model for soybean cropping system under different soils, in the Indian agro-ecosystem, predicted that the potential recharge ranged from 2.1 to 23.3% of monsoon season rainfall. Simulated potential GR from water harvesting structures by the IPGRS model ranged from 83.0 to 90.2% of accumulated surface runoff in the water harvesting structures. Modeled surface runoff for the cropped and ravine land-use system of an agro-ecosystem ranged from 10 to 20% of the rainfall. Simulated annual streamflow of the Brahmani River in India using the PRMS and multi-model ensemble climate change scenarios indicated an increase in magnitude and frequency of high (flood) flows, and a decrease in low flow along with the greater frequency of low flows. The USLE has been successfully applied in a semi-arid and sub-humid climate of India for different cropping systems, and the long-term average annual soil loss ranged from 0.06 to 0.41, and 1.26 to 4.03 $Mg\,ha^{-1}\,yr^{-1}$, respectively. Modeling studies on some of the water-regulating agro-ecosystem services showed simulated climate change effects vary across models and the land use regions within the agro-ecosystem of a country. These water-regulating agro-ecosystem models provide valuable information and policy decision for preparedness, adaptive planning, and measures to mitigate the climate change effects on water-regulating agro-ecosystem services.

Abbreviations

BMPs, best management practices; BREB, Bowen ratio energy balance; CMIP3, Coupled Model Inter-comparison Project phase 3; CREAMS, Chemicals, Runoff, and Erosion from Agricultural Management Systems; ET, evapotranspiration; ET_0, reference evapotranspiration; FAO, Food and Agriculture Organization; GA, Green–Ampt; GCM, General Circulation Model; GR, groundwater recharge; HEC-HMS, Hydrologic Engineering Center–Hydrologic Modeling System; HWDS, holistic water depth simulation; ICID, International Commission on Irrigation and Drainage; IPGRS, integrated potential groundwater recharge simulation; KINEROS, Kinematic Runoff and Erosion Model; MEA, Millennium Ecosystem Assessment; MMF, Morgan–Morgan–Finney; MPRE, maximum percent relative error; MT, mass transfer; NAPI, normalized antecedent precipitation index; PB, percent bias; PE, pan evaporation; PM, Penman–Monteith; PRMS, Precipitation Runoff Modeling Systems; PT, Priestley–Taylor; RCPs, representative concentration pathways; RPs, recharge ponds; RUSLE, revised universal soil loss equation; SCS-CN, Soil Conservation Service Curve Number; SPAW, soil–plant–air–water; SVAT, soil–vegetation–atmosphere transfer; SWAT, Soil Water Assessment Tool; USLE, universal soil loss equation.

References

Abeysingha, N. S., Islam, A., & Singh, M. (2018). Assessment of climate change impact on flow regimes over the Gomti River basin under IPCC AR5 climate change scenarios. *Water & Climate Change, 11*, 303–326. https://doi.org/10.2166/wcc.2018.039

Abeysingha, N. S., Singh, M., Islam, A., & Sehgal, V. K. (2016). Climate change impacts on irrigated rice and wheat production in Gomti River basin of India: A case study. *SpringerPlus, 5*, 1250. https://doi.org/10.1186/s40064-016-2905-y

Abtew, W. (2001). Evaporation estimation for Lake Okeechobee in South Florida. *Journal of Irrigation and Drainage Engineering, 127*, 140–146.

Ali, S. (2016). Estimation of time variant water availability and irrigation potential of small ponds in a semi-arid region of Rajasthan, India. *Journal of the Institution of Engineers (India) Series A, 97*, 43–51. https://doi.org/10.1007/s40030-016-0141-7

Ali, S., & Ghosh, N. C. (2016). A methodology for estimation of wetting front length and potential recharge under variable depth of ponding. *Journal of Irrigation and Drainage Engineering, 142*, 04015027.

Ali, S., & Ghosh, N. C. (2019). Integrated modelling of potential recharge in small recharge ponds. *Arabian Journal of Geosciences, 12*, 455. https://doi.org/10.1007/s12517-019-4634-3

Ali, S., Ghosh, N. C., Mishra, P. K., & Singh, R. K. (2015). A holistic water depth simulation model for small ponds. *Journal of Hydrology, 529*, 1464–1477. https://doi.org/10.1016/j.jhydrol.2015.08.035

Ali, S., Ghosh, N. C., & Singh, R. (2008). Evaluation of evaporation estimate model for water surface evaporation in semi-arid region, India. *Hydrological Processes, 22*, 1093–1106.

Ali, S., Ghosh, N. C., & Singh, R. (2010). Rainfall-runoff simulation using normalized antecedent precipitation index. *Hydrological Sciences Journal, 55*, 266–274. https://doi.org/10.1080/02626660903546175

Ali, S., Ghosh, N. C., Singh, R., & Sethy, B. K. (2013). Generalized explicit models for estimation of wetting front length and potential recharge. *Water Resources Management, 27*, 2429–2445.

Ali, S., & Islam, A. (2018). Solution to Green–Ampt infiltration model using a two-step curve-fitting approach. *Environmental Earth Sciences, 77*, –271. https://doi.org/10.1007/s12665-018-7449-8

Ali, S., Islam, A., Mishra, P. K., & Sikka, A. K. (2016a). Green–Ampt approximations: A comprehensive analysis. *Journal of Hydrology, 535*, 340–355.

Ali, S., Mishra, P. K., Islam, A., & Alam, N. M. (2016b). Simulation of water temperature in small pond using parametric statistical models: Implications of climate warming. *Journal of Environmental Engineering, 142*, 04015085. https://doi.org/10.1061/(ASCE)EE.1943-7870.0001050

Ali, S., Samra, J. S., Singh, K. D., & Prasad, S. N. (2002). Soil loss estimation model for South-Eastern Rajasthan. *Journal of the Institution of Engineers (India) AG, 83*, 1–4.

Ali, S., Sethy, B. K., Singh, R. K., Parandiyal, A. K., & Kumar, A. (2017). Quantification of hydrologic response of staggered contour trenching for horti-pastoral land use system in small ravine watersheds: A paired watershed approach. *Land Degradation and Development, 28*, 1237–1252.

Ali, S., & Sharda, V. N. (2005). Evaluation of the universal soil loss equation (USLE) in semi-arid and subhumid climates of India. *Applied Engineering in Agriculture, 21*, 217–225. https://doi.org/10.13031/2013.18156

Ali, S., & Singh, R. (2001). Regional runoff frequency curve for small watersheds in the Hirakud catchment of eastern India. *Applied Engineering in Agriculture, 17*, 285–292.

Alison, G. P. (2010). Ecosystem services and agriculture: Tradeoffs and synergies. *Philosophical Transactions of the Royal Society B, 365*, 2959–2971. https://doi.org/10.1098/rstb.2010.0143

Allen, L. H. (1990). Plant response to rising carbon dioxide and potential interaction with air pollutants. *Journal of Environmental Quality, 19*, 15–34. https://doi.org/10.2134/jeq1990.00472425001900010002x

Allen, R. G., Pereira, L. S., Raes, D., & Smith, M. (1998). *Crop evapotranspiration: Guidelines for computing crop water requirements.* FAO Irrigation and Drainage Paper 56. United Nations FAO.

Altunkaynak, A. (2007). Forecasting surface water level fluctuations of Lake Van by artificial neural networks. *Water Resources Management, 21*, 399–408.

Arnold, J. G., Srinivasan, R., Muttiah, S., & Williams, J. R. (1998). Large-area hydrologic modeling and assessment: Part I. Model development. *Journal of the American Water Resources Association, 34*, 73–89.

Arora, V. (2002). Modeling vegetation as a dynamic component in soil–vegetation–atmosphere transfer schemes and hydrological models. *Reviews of Geophysics, 40*, 1006. https://doi.org/10.1029/2001RG000103

Bangash, R. F., Passuello, A., Sanchez-Canales, M., Terrado, M., López, A., Elorza, F. J., Ziv, G., Acuña, V., & Schuhmacher, M. (2013). Ecosystem services in Mediterranean river basin: Climate change impact on water provisioning and erosion control. *The Science of the Total Environment, 458–460*, 246–255. https://doi.org/10.1016/j.scitotenv.2013.04.025

Beven, K., & Germann, P. (2013). Macropores and water flow in soils revisited. *Water Resources Research, 49*, 3071–3092. https://doi.org/10.1002/wrcr.20156

Bhuvaneswari, K., Geethalakshmi, V., Lakshmanan, A., Srinivasan, R., & Sekhar, N. U. (2013). The impact of El Niño/southern oscillation on hydrology and rice productivity in the Cauvery basin in India: Application of the soil and water assessment tool. *Weather and Climate Extremes, 2*, 39–47. https://doi.org/10.1016/j.wace.2013.10.003

Biswas, S. S., & Pani, P. (2015). Estimation of soil erosion using RUSLE and GIS techniques: A case study of Barakar River basin, Jharkhand, India. *Modeling Earth Systems and Environment, 1*, 42. https://doi.org/10.1007/s40808-015-0040-3

Brocca, L., Melone, F., & Moramarco, T. (2008). On the estimation of antecedent wetness conditions in rainfall-runoff modeling. *Hydrological Processes, 22*, 629–642.

Chen, J., Cui, T., Wang, H., Liu, G., Gilfedder, M., & Bai, Y. (2018). Spatio-temporal evolution of water-related ecosystem services: Taihu Basin. *Peer J, 6*, e5041. https://doi.org/10.7717/peerj.5041

Chittarpur, B. M., & Patil, D. K. (2017). Ecosystem services render by tree based land use system. *Indian Journal of Agricultural Sciences, 87*, 1419–1429.

Cochrane, T. A., & Flanagan, D. C. (1999). Assessing water erosion in small watersheds using WEPP with GIS and digital elevation models. *Journal of Soil and Water Conservation, 54*, 678–685.

Dabral, P. P., Baithuri, N., & Pandey, A. (2008). Soil erosion assessment in a hilly catchment of north eastern India using USLE, GIS and remote sensing. *Water Resources Management, 22*, 1783–1798.

Dagadu, J. S., & Nimbalkar, P. T. (2012). Infiltration studies of different soils under different soil conditions and comparison of infiltration models with field data. *International Journal of Advanced Engineering Technology, 3*, 154–157.

Daily, G. C., & Matson, P. A. (2008). Ecosystem services: From theory to implementation. *Proceedings of the National Academy of Sciences of the USA, 105*, 9455–9456. https://doi.org/10.1073/pnas.0804960105

Dhruva Narayana, V. V., & Babu, R. (1983). Estimation of soil erosion in India. *Journal of Irrigation and Drainage Engineering, 109*, 419–434.

Dominati, E., Patterson, M., & Mackay, A. (2010). A framework for classifying and quantifying the natural capital and ecosystem services of soils. *Ecological Economics, 69*, 1858–1868. https://doi.org/10.1016/j.ecolecon.2010.05.002

Dutta, V., Kumar, R., & Sharma, U. (2015). Assessment of human-induced impacts on hydrological regime of Gomti river basin, India. *Management of Environmental Quality, 26*, 631–649. https://doi.org/10.1108/MEQ-11-2014-0160

Elliott, J., Kelly, D., Chryssanthacopoulos, J., Glotter, M., Jhunjhnuwala, K., Best, N., Wilde, M., & Foster, I. (2014). The parallel system for integrating impact models and sectors (pSIMS). *Environmental Modelling & Software, 62*, 509–516. https://doi.org/10.1016/j.envsoft.2014.04.008

Falkenmark, M. (2003). Freshwater as shared between society and ecosystems: From divided approaches to integrated challenges. *Philosophical Transactions of the Royal Society B, 358*, 2037–2049. https://doi.org/10.1098/rstb.2003.1386

FAO. (2018). *Arable land.* https://data.worldbank.org/indicator/AG.LND.ARBL.HA.PC

Feldman, A. D. (1981). HEC models for water resources simulation: Theory and experience. In V. T. Chow (Ed.), *Advances in hydroscience, Vol. 12* (pp. 297–423). Academic Press.

Flint, A. L., & Ellett, K. M. (2004). The role of the unsaturated zone in artificial recharge at San Gorgonio Pass, California. *Vadoze Zone Journal, 3*, 763–774. https://doi.org/10.2136/vzj2004.0763

Francesconi, W., Srinivasan, R., Pérez-Miñana, E., Willcock, S. P., & Quintero, M. (2016). Using the soil and water assessment tool (SWAT) to model ecosystem services: A systematic review. *Journal of Hydrology, 535*, 625–636.

Gordon, L. J., Finlayson, C. M., & Falkenmark, M. (2010). Managing water in agriculture for food production and other ecosystem services. *Agricultural Water Management, 97*, 512–519. https://doi.org/10.1016/j.agwat.2009.03.017

Gosain, A. K., Rao, S., & Arora, A. (2011). Climate change impact assessment of water resources of India. *Current Science, 101*, 356–371.

Green, W. H., & Ampt, G. (1911). Studies on soil physics. *The Journal of Agricultural Science, 4*, 1–24. https://doi.org/10.1017/S0021859600001441

Gundalia, M. (2018). Estimation of infiltration rate based on complementary error function peak for Ozat watershed in Gujarat (India). *International Journal of Hydrology, 2*, 289–294. https://doi.org/10.15406/ijh.2018.02.00083

Gupta, S., & Kumar, S. (2017). Simulating climate change impact on soil erosion using RUSLE model: A case study in a watershed of mid-Himalayan landscape. *Journal of Earth System Science, 126*, 43. https://doi.org/10.1007/s12040-017-0823-1

Guswa, A. J., Brauman, K. A., Brown, C., Hamel, P., Keeler, B. L., & Sayre, S. S. (2014). Ecosystem services: Challenges and opportunities for hydrologic modeling to support decision making. *Water Resources Research, 50*, 4535–4544.

Haan, C. T., Allred, T. B., Storm, D. E., Sabbagh, G. J., & Prabhu, S. (1995). A statistical procedure for evaluating hydrologic/water quality models. *Transactions of the American Society of Agricultural Engineers, 38*, 725–733.

Harbeck, G. E. (1962). A practical field technique for measuring reservoir evaporation utilizing mass-transfer theory. *US Geological Survey Professional Paper, 272-E*, 101–105.

Holtan, H. N. (1961). *A concept of infiltration estimates in watershed engineering, Rep. ARS41-51.* USDA-ARS.

Horton, R. (1933). The role infiltration in the hydrologic cycle. *Transactions, American Geophysical Union, 14*, 446–460. https://doi.org/10.1029/TR014i001p00446

Hsiao, J., Swann, A. L. S., & Kim, S.-H. (2019). Maize yield under a changing climate: The hidden role of vapor pressure deficit. *Agricultural and Forest Meteorology, 279*, 107692.

Irmak, S., Payero, J. O., Martin, D. L., Irmak, A., & Howell, T. A. (2006). Sensitivity analyses and sensitivity coefficients of standardized daily ASCE Penman–Monteith equation. *Journal of Irrigation and Drainage Engineering, 132*, 564–578.

Islam, A., Ahuja, L. R., Garcia, L. A., Ma, L., & Saseesndran, A. S. (2012a). Modeling the effect of elevated CO_2 and climate change on potential evapotranspiration in the semi-arid central Great Plains. *Transactions of the American Society of Agricultural and Biological Engineers, 55*, 2135–2146. https://doi.org/10.13031/2013.42505

Islam, A., Shirsath, P. B., Kumar, S. N., Subhash, N., Sikka, A. K., & Aggarwal, P. K. (2014). Modeling water management and food security in India under climate change. In L. R. Ahuja, L. Ma, & R. J. Lascano (Eds.), *Practical applications of agricultural system models to optimize the use of limited water* (pp. 267–316). ASA, SSSA, and CSSA. https://doi.org/10.2134/advagricsystmodel5.c11

Islam, A., Sikka, A. K., Saha, B., & Singh, A. (2012b). Streamflow response to climate change in the Brahmani River Basin, India. *Water Resources Management, 26*, 1409–1424.

Jejurkar, C. L., & Rajurkar, M. P. (2014). An investigational approach for the modelling of infiltration process in a clay soil. *KSCE Journal of Civil Engineering, 2*, 72–76. https://doi.org/10.1007/s12205-014-0149-3

Jones, J. W., Antle, J. M., Basso, B. O., Boote, K. J., Conant, R. T., Foster, I., Godfray, H. C. J., Herrero, M., Howitt, R. E., Janssen, S., Keating, B. A., Munoz-Carpena, R., Porter, C. H., Rosenzweig, C., & Wheeler, T. R. (2016). Brief history of agricultural systems modeling. *Agricultural Systems, 155*, 240–254. https://doi.org/10.1016/j.agsy.2016.05.014

Kakahaji, H., Banadaki, H. D., Kakahaji, A., & Kakahaji, A. (2013). Prediction of Urmia Lake water level fluctuations by using analytical, linear statistic and intelligent methods. *Water Resources Management, 27*, 4469–4492. https://doi.org/10.1007/s11269-013-0420-2

Khare, D., Mondal, A., Kundu, S., & Mishra, P. K. (2016). Climate change impact on soil erosion in the Mandakini River basin, North India. *Applied Water Science*, *7*, 2373–2383. https://doi.org/10.1007/s13201-016-0419-y

Khatal, S. A., Ali, S., Hasan, M., Singh, D. K., Mishra, A. K., & Iquebal, M. A. (2018). Assessment of groundwater recharge in a small ravine watershed in semi-arid region of India. *International Journal of Current Microbiology and Applied Sciences*, *7*, 2552–2565.

Kim, H. J., Sidle, R. C., & Moore, R. D. (2005). Shallow lateral flow from a forested hillslope: Influence of antecedent wetness. *Catena*, *60*, 293–306. https://doi.org/10.20546/ijcmas.2018.702.311

Kimball, B. A. (2007). Global changes and water resources. In R. J. Lascano & R. E. Sojka (Eds.), *Irrigation of agricultural crops* (2nd ed., pp. 627–653)). ASA, CSSA, and SSSA. https://doi.org/10.2134/agronmonogr30.2ed.c17

Kingston, D. G., Todd, M. C., Taylor, R. G., Thompson, J. R., & Arnell, N. W. (2009). Uncertainty in the estimation of potential evapotranspiration under climate change. *Geophysical Research Letters*, *36*, L20403. https://doi.org/10.1029/2009GL040267

Kirschbaum, M. U. F., & McMillan, A. M. S. (2018). Warming and elevated CO_2 have opposing influences on transpiration. Which is more important? *Current Forestry Reports*, *4*, 51–71. https://doi.org/10.1007/s40725-018-0073-8

Knisel, W. G. (1980). *CREAMS: A field scale model for chemicals, runoff and erosion from agricultural management systems*. Conservation Research Report no. 26. USDA-SEA.

Kostiakov, A. N. (1932). The dynamics of the coefficient of water percolation in soils and the necessity for studying it from a dynamic point of view for purpose of amelioration. *Transactions of 6th Committee International Society of Soil Science, Russia, Part A*, 15–21.

Kumar, S. R., Patwari, B. C., & Bhuniya, P. K. (1995). *Infiltration studies: Dudhnai sub-basin (Assam and Meghalaya)*. Technical Report CS (AR)-183. National Institute of Hydrology. http://117.252.14.250:8080/xmlui/handle/123456789/161

Lakshmanan, A., Geethalakshmi, V., Srinivasan, R., Sekhar, N. U., & Annamalai, H. (2011). Climate change adaptation strategies in Bhavani basin using swat model. *Applied Engineering in Agriculture*, *27*, 887–893. https://doi.org/10.13031/2013.40623

Lal, R. (Ed.) (1999). *Soil quality and soil erosion*. Soil and Water Conservation Society.

Lal, R. (2003). Soil erosion and the global carbon budget. *Environment International*, *29*, 437–450. https://doi.org/10.1016/S0160-4120(02)00192-7

Lal, R., Ahmadi, M., & Bajracharya, R. M. (2000). Erosional impacts on soil properties and corn yield on Alfisols in Central Ohio. *Land Degradation and Development*, *11*, 575–585.

Leavesley, G. H., Markstrom, S. L., Restrepo, P. J., & Viger, R. J. (2002). A modular approach to addressing model design, scale, and parameter estimation issues in distributed hydrological modelling. *Hydrological Processes*, *16*, 173–187.

Longbottom, T. L., Townsend-Small, A., Owen, L. A., & Murari, M. K. (2014). Climatic and topographic controls on soil organic matter storage and dynamics in the Indian Himalaya: Potential carbon cycle–climate change feedbacks. *Catena*, *119*, 125–135. https://doi.org/10.1016/j.catena.2014.03.002

Machiwal, D., Jha, M. K., & Mal, B. C. (2006). Modelling infiltration and quantifying spatial soil variability in a watershed of Kharagpur, India. *Biosystems Engineering*, *95*, 569–582. https://doi.org/10.1016/j.biosystemseng.2006.08.007

Maeda, E. E., Pellikka, P. K. E., Siljander, M., & Clark, B. J. E. (2010). Potential impacts of agricultural expansion and climate change on soil erosion in the eastern Arc Mountains of Kenya. *Geomorphology*, *123*, 279–289.

Mahapatra, S., Jha, M. K., Biswal, S., & Senapati, D. (2020). Assessing variability of infiltration characteristics and reliability of infiltration models in a tropical sub-humid region of India. *Scientific Reports*, *10*, 1515. https://doi.org/10.1038/s41598-020-58333-8

Martin, P., Rosenberg, N. J., & McKenney, M. S. (1989). Sensitivity of evapotranspiration in a wheat field, a forest, and grassland to changes in climate and direct effects of carbon dioxide. *Climate Change*, *14*, 117–151.

Matson, P. A., Parton, W. J., Power, A. G., & Swift, M. J. (1997). Agricultural intensification and ecosystem properties. *Science*, *277*, 504–509.

MEA (Millennium Ecosystem Assessment). (2005). Millennium ecosystem assessment. In *Ecosystems and human well-being: Biodiversity synthesis*. World Resources Institute.

Mishra, V., & Lilhare, R. (2016). Hydrologic sensitivity of Indian sub-continental river basins to climate change. *Global and Planetary Change*, *139*, 78–96. https://doi.org/10.1016/j.gloplacha.2016.01.003

Mondal, A., Khare, D., & Kundu, S. (2016). Impact assessment of climate change on future soil erosion and SOC loss. *Natural Hazards, 82,* 1515–1539. https://doi.org/https://doi.org/10.1007/s11069-016-2255-7

Mondal, A., Khare, D., Kundu, S., Meena, P. K., Mishra, P. K., & Shukla, R. (2014a). Impact of climate change on future soil erosion in different slope, land use and soil type conditions in a part of Narmada river basin. *Journal of Hydrologic Engineering, 20,* C5014003-1–C5014003-12. https://doi.org/10.1061/(ASCE)HE.1943-5584.0001065

Mondal, A., Khare, D., Kundu, S., & Mishra, P. K. (2014b). Detection of land use change and future prediction with Markov chain model in a part of Narmada River basin, Madhya Pradesh. In M. Singh, R. B. Singh, & M. I. Hassan (Eds.), *Landscape Ecology and Water Management, Proceeding of IGU Rohtak Conference, Vol. 2* (pp. 3–14). Springer.

Mullan, D. (2013). Managing soil erosion in Northern Ireland: A review of past and present approaches. *Agriculture, 3,* 684–699. https://doi.org/10.3390/agriculture3040684

Nagamani, K., & Mariappan, N. V. E. (2017). Remote sensing, GIS and crop simulation models: A review. *International Journal of Current Research in Biosciences and Plant Biology, 4,* 80–92. https://doi.org/10.20546/ijcrbp.2017.408.011

Narsimlu, B., Gosain, A. K., & Chahar, B. R. (2013). Assessment of future climate change impacts on water resources of upper Sind River Basin, India using SWAT model. *Water Resources Management, 27,* 3647–3662.

Nearing, M. A., Pruski, F. F., & O'Neal, M. R. (2004). Expected climate change impacts on soil erosion rates: A review. *Journal of Soil and Water Conservation, 59,* 43–50.

Pandey, A., Mathur, A., Mishra, S. K., & Mal, B. C. (2009). Soil erosion modeling of a Himalayan watershed using RS and GIS. *Environmental Earth Sciences, 59,* 399–410.

Passcheir, R. H. (1996). *Evaluation of hydrologic model packages.* Technical Report Q2044. WL/Delft Hydraulics.

Pathak, H., Chakrabarti, B., Mina, U., Pramanik, P., & Sharma, D. K. (2017). Ecosystem service of wheat (*Triticum aestivum*) production with conventional and conservation agricultural practices in the indo-Gangatic plains. *Indian Journal of Agricultural Sciences, 87,* 3–7.

Patle, G. T., Sikar, T. T., Rawat, K. S., & Singh, S. K. (2019). Estimation of infiltration rate from soil properties using regression model for cultivated land. *Geology, Ecology, and Landscapes, 3,* 1–13. https://doi.org/10.1080/24749508.2018.1481633

Penman, H. L. (1948). Natural evaporation from open water, bare soil and grass. *Proceedings of the Royal Society A, 193,* 120–145. https://doi.org/10.1098/rspa.1948.0037

Philip, J. R. (1957). Theory of infiltration: Sorptivity and algebraic equations. *Soil Science, 84,* 257–265.

Pimentel, D. (2006). Soil erosion: A food and environmental threat. *Environment, Development and Sustainability, 8,* 119–137. https://doi.org/10.1007/s10668-005-1262-8

Pimental, D., & Sparks, D. L. (2000). Soil as an endangered escosystem. *BioScience, 50,* 947. https://doi.org/10.1641/0006-3568(2000)050[0947:SAAEE]2.0.CO;2

Priestley, C. H. B., & Taylor, R. J. (1972). On the assessment of surface heat flux and evaporation using large scale parameters. *Monthly Weather Review, 100,* 81–92. https://doi.org/10.1175/1520-0493(1972)100<0081:OTAOSH>2.3.CO;2

Priya, A., Islam, A., Nema, A., & Sikka, A. K. (2015). Assessing sensitivity of reference evapotranspiration to changes in climatic variables: A case study of Akola, India. *MAUSAM, 66,* 777–784.

Priya, A., Nema, A., & Islam, A. (2014). Effect of climate change and elevated co_2 on reference evapotranspiration in Varanasi, India: A case study. *Journal of Agrometeorology, 16,* 44–51.

Qi, S., Sun, G., Wang, Y., McNulty, S. G., & Moore Myers, J. A. (2009). Streamflow response to climate and land use changes in a coastal watershed in North Carolina. *Transactions of the American Society of Agricultural and Biological Engineers, 52,* 739–749.

Ramirez, J. A., & Finnerty, B. (1996). CO_2 and temperature effects on evapotranspiration and irrigated agriculture. *Journal of Irrigation and Drainage Engineering, 122,* 155–163.

Rasool, T., Dar, A. Q., & Wani, M. A. (2021a). Comparative evaluation of infiltration models under different land covers. *Water Resources, 48,* 624–634. https://doi.org/10.1134/S0097807821040175

Rasool, T., Dar, A. Q., & Wani, M. A. (2021b). Comparison of infiltration model parameter estimation techniques under different land use/land covers. *International Journal of Hydrology Science and Technology, 12,* 448–476. https://doi.org/10.1504/IJHST.2021.118308

Richards, L. A. (1931). Capillary conduction of liquids through porous mediums. *Journal of Applied Physics, 1,* 318–333. https://doi.org/10.1063/1.1745010

Robertson, G. P., Gross, K. L., Hamilton, S. K., Landis, D. A., Schmidt, T. M., Snapp, S. S., & Swinton, S. M. (2014). Farming for ecosystem services: An ecological approach to production agriculture. *BioScience, 64*, 404–415. https://doi.org/10.1093/biosci/biu037

Rockström, J., Folke, C., Gordon, L., Hatibu, N., Jewitt, G., Penning de Vries, F., Rwehumbiza, F., Sally, H., Savenije, H., & Schulze, R. (2004). A watershed approach to upgrade rainfed agriculture in water scarce regions through water system innovations: An integrated research initiative on water for food and rural livelihoods in balance with ecosystem functions. *Physics and Chemistry of the Earth, Parts A/B/C, 29*, 1109–1118. https://doi.org/10.1016/j.pce.2004.09.016

Roy, B. P., & Singh, H. (1995). *Infiltration study of a sub-basin.* Case Study CS (AR)-170. National Institute of Hydrology.

Rushton, K. (1997). Recharge from permanent water bodies. In I. Simmers (Ed.), *Recharge of phreatic aquifers in semi arid areas* (pp. 215–255). A Baalkema.

Sacks, L. A., Lee, T. M., & Radell, M. J. (1994). Comparison of energy-budget evaporation. Losses from two morphologically different Florida seepage lakes. *Journal of Hydrology, 156*, 311–334.

Sahle, M., Saito, O., Fürst, C., & Yeshitela, K. (2019). Quantifying and mapping of water-related ecosystem services for enhancing the security of the food-water-energy nexus in tropical data-sparse catchment. *The Science of the Total Environment, 646*, 573–586.

Saxton, K. E., & Willey, P. H. (2006). The SPAW model for agricultural field and pond hydrologic simulation. In V. P. Singh & D. K. Frevert (Eds.), *Watershed models* (pp. 401–435). CRC Press.

Scanlon, B. R., Mukherjee, A., Gates, J., Reedy, R. C., & Sinha, A. K. (2010). Groundwater recharge in natural dune systems and agricultural ecosystems in the Thar Desert region, Rajasthan, India. *Hydrogeology Journal, 18*, 959–972. https://doi.org/. https://doi.org/10.1007/s10040-009-0555-7

Sharda, V. N., & Ali, S. (2008). Evaluation of the universal soil loss equation in semi-arid and sub-humid climates of India using stage dependent C-factor. *Indian Journal of Agricultural Sciences, 78*, 422–427.

Sharda, V. N., Dogra, P., & Prakash, C. (2010). Assessment of production losses due to water erosion in rainfed areas of India. *Journal of Soil and Water Conservation, 65*, 79–91.

Sharda, V. N., & Ojasvi, P. R. (2016). A revised soil erosion budget for India: Role of reservoir sedimentation and land-use protection measures. *Earth Surface Processes and Landforms, 41*, 2007–2023. https://doi.org/10.1002/esp.3965

Sikka, A. K., Islam, A., & Rao, K. V. (2018). Climate-smart land and water management for sustainable agriculture. *Irrigation and Drainage, 67*, 72–81. https://doi.org/10.1002/ird.2162

Singh, V. P., & Woolhiser, D. A. (2002). Mathematical modeling of watershed hydrology. *Journal of Hydrologic Engineering, 7*, 270–292. https://doi.org/10.1061/(ASCE)1084-0699(2002)7:4(270)

Singh, V. P., & Xu, C. Y. (1997). Evaluation and generalization of 13 mass transfer equations for determining free water evaporation. *Hydrological Processes, 11*, 311–323.

Smukler, S. M., Philpott, S. M., Jackson, L. E., Klein, A. M., DeClerck, F., Winowiecki, L., & Palm, C. A. (2012). Ecosystem services in agricultural landscapes. In J. C. Ingram, F. DeClerck, & C. R. Rio (Eds.), *Integrating ecology and poverty reduction* (pp. 17–51). Springer Science + Business Media, LLC.

Tang, J. L., Cheng, X. Q., Zhu, B., Gao, M. R., Wang, T., Zhang, X. F., Zhao, P., & You, X. (2015). Rainfall and tillage impacts on soil erosion of sloping cropland with subtropical monsoon climate: A case study in hilly purple soil area, China. *Journal of Mountain Science, 12*, 134–144. https://doi.org/10.1007/s11629-014-3241-8

Tejpal (2013). Relief analysis of the Tangri watershed in the lower Shivalik and Piedmont zone of Haryana and Punjab. *Indian Journal of Spatial Science, 4*, 9–20.

Turner, B. L., Menendez, H. M., Gates, R., Tedeschi, L. O., & Atzori, A. S. (2016). System dynamics modeling for agricultural and natural resource management issues: Review of some past cases and forecasting future roles. *Resources, 5*, 40. https://doi.org/10.3390/resources5040040

USDA-SCS. (1993). Estimation of direct runoff from storm runoff. In *Hydrology–National Engineering Handbook* (pp. 2–18) USDA-SCS.

Vandana, K., Islam, A., Sarthi, P. P., Sikka, A. K., & Kapil, H. (2018). Assessment of potential impact of climate change on streamflow: A case study of the Brahmani River Basin, India. *Water & Climate Change, 10*, 624–641. https://doi.org/10.2166/wcc.2018.129

Vatankhah, A. R. (2015). Discussion of "Modified Green–Ampt infiltration model for steady rainfall" by J. Almedeij and I. I. Esen. *Journal of Hydrologic Engineering, 20*, 07014011. https://doi.org/10.1061/(ASCE)HE.1943-5584.0001110

Vereecken, H., Schnepf, A., Hopmans, J. W., Javaux, M., Or, D., Roose, T., Vanderborght, J., Young, M. H., Amelung, W., Aitkenhead, M., Allison, S. D., Assouline, S., Baveye, P., Berli, M., Rüggemann, N., Finke, P., Flury, M., Gaiser, T., Govers, G., . . . Young, I. M. (2016). Modeling soil processes: Review, key challenges, and new perspectives. *Vadose Zone Journal, 15,* 1–57. https://doi.org/10.2136/vzj2015.09.0131

Vigerstol, K. L., & Aukema, J. E. (2011). A comparison of tools for modeling freshwater ecosystem services. *Journal of Environmental Management, 92,* 2403–2409. https://doi.org/10.1016/j.jenvman.2011.06.040

Wang, X. K., Lu, W. Z., Cao, S. Y., & Fang, D. (2007). Using time-delay neural network combined with genetic algorithm to predict runoff level of Linshan watershed, Sichuan, China. *Journal of Hydrologic Engineering, 12,* 231–236. https://doi.org/10.1061/(ASCE)1084-0699(2007)12:2(231)

Woolhiser, D. A., Smith, R. E., & Goodrich, D. C. (1990). *KINEROS, A Kinematic Runoff and Erosion Model.* Documentation and user manual, ARS-77. USDA-ARS.

Wu, Y., Liu, S., & Abdul-Aziz, O. I. (2011). Hydrological effects of the increased CO_2 and climate change in the upper Mississippi River basin using a modified SWAT. *Climate Change, 110,* 977–1003.

Zhang, Y., Hernasndez, M., Anson, E., Nearing, M. A., Wei, H., Stone, J. J., & Heilman, R. (2012). Modeling climate change effects on runoff and soil erosion in southeastern Arizona rangelands and implications for mitigation with conservation practices. *Journal of Soil and Water Conservation, 67,* 390–405.

6

Xiaoyu Gao, Zailin Huo,
Zhongyi Qu, and Pengcheng Tang

Modeling Agricultural Hydrology and Water Productivity to Enhance Water Management in the Arid Irrigation District of China

Abstract

Capillary rise from shallow groundwater can decrease the need for irrigation water. However, simple techniques do not exist to quantify the contribution of capillary flux to crop water use. In this study we develop the Agricultural Water Productivity Model for Shallow Groundwater (AWPM-SG) for calculating capillary fluxes from shallow groundwater using readily available data. The model combines an analytical solution of upward flux from groundwater with the Environmental Policy Integrated Climate (EPIC) crop growth model. The AWPM-SG model was calibrated and validated with a 2-yr lysimetric experiment with maize (*Zea mays* L.). Predicted soil moisture, groundwater depth, and leaf area index agreed with the observations. Based on the calibrated AWPM-SG model, a 5-yr regional water productivity (WP) and water budgets assessment was performed. The results showed that the groundwater contribution to crop evapotranspiration (ET) would be up to 65% with a groundwater depth of 1.0–1.5 m, but the agricultural productivity would be relatively low resulting from a waterlogged root zone. Additionally, deep groundwater could result in a reduced WP due to less capillary rise, whereas WP would be 20.2 and 19.8 kg ha^{-1} mm^{-1} with groundwater depth of 2.5–3.0 and 3.0–4.5 m under irrigation amount of 100–300 mm. Furthermore, limited irrigation can enhance the contribution of groundwater to WP and irrigation water productivity,

---

Enhancing Agricultural Research and Precision Management for Subsistence Farming by Integrating System Models with Experiments, First Edition. Edited by Dennis J. Timlin and Saseendran S. Anapalli.

DOI: 10.1002/9780891183891.ch06

which is significant with groundwater depth increasing. Thus, at the regional scale, the spatial distribution of groundwater levels needs to be considered for making irrigation decisions.

Introduction

Water scarcity has become a serious limiting factor for worldwide agricultural and economic development with increasing food demands and populations. However, compared with developed countries, water productivity (WP) is low in most developing countries. In many semi-arid and arid regions of the world, irrigated agriculture accounts for 90% of the total water use. Improving agricultural water productivity is a priority for ensuring water and food security (Dalin et al., 2015).

Agricultural water saving directly changes the soil water content of the root zone and groundwater upward flux, which causes changes to the agricultural water cycle and crop growth (Bouman, 2007; Morison et al., 2008; Jaksa & Sridhar, 2015). Agricultural water cycles are the driving processes for agricultural water productivity, and the complex relationships between crop water consumption, soil water content, groundwater, and irrigation at the regional scale are still unclear. Therefore, based on understanding the linked process of crop growth and agricultural water cycles, quantifying the effect of agricultural water saving on agricultural water productivity is the basis for realizing effective water use.

Especially in a shallow groundwater district, like the Hetao irrigation district, groundwater levels apparently fluctuate as a result of irrigation over the crop growing period (Xu et al., 2015) and the agricultural water cycle becomes more complex. With the implementation of water saving strategies, the exchange between soil water and groundwater will be changing with declining groundwater. Furthermore, the groundwater upward flux in the water table needs quantification because it can be used to supplement surface water irrigation and in closed basins can possibly save water for irrigating additional areas (Huo et al., 2012; Yang et al., 2007).

Several experiments have been carried out to quantify the amount of water that is used by agricultural crops from groundwater. Bargahei and Mosavi (2006) reported in an experimental study that groundwater supplemented between 53 and 55% of the water of safflower utilizing saline groundwater and about 82% for safflower utilized fresh groundwater at groundwater depths (between 50 and 120 cm). Ghamarnia found that two-thirds of the average annual safflower (*Carthamus tinctorius* L.) water requirement (65 cm) could be met with groundwater table depth from 0.6 to 0.8 m (Ghamarnia et al., 2013). With lysimeter experiments, Kahlown et al. (2005) reported that the wheat (*Triticum aestivum* L.) crop did not need any supplementary irrigation with a water table at a 0.5-m depth and 80% sunflower's water need was supplied by groundwater at a 0.5-m depth.

However, due to their inconvenience and expensive maintenance of field experimental methods, the processes-based models are preferred to quantify the capillary rise. The groundwater–soil–plant–atmosphere continuous (GSPAC) system is important to understand the agricultural water cycle (Li et al., 2011;

Wang et al., 2004; Zhou et al., 2005). Numerical models such as HYDRUS (Simunek et al., 2005) and soil–water–atmosphere–plant (SWAP) (van Dam et al., 2008) require many soil and crop parameters that limit the use of models to some extent (Schoups et al., 2005). In contrast, conceptual models of agricultural water cycles based on the water balance method are widely used due to their simple structure and limited parameters. Considering the impact of soil water on the crop evapotranspiration (ET), some studies incorporated a root uptake model in the water balance model (Kendy et al., 2003).

Agricultural water productivity is not only related to the water cycle process but also to the crop growing process. In agricultural water management, the relationship between crop yield and crop ET is often characterized using a water production function based on abundant field experiments, but it ignores the formation process of crop growing, development, final biomass and yield, and ET. Therefore, crop models become important to quantify the agricultural water productivity (WP). Models such as Simple and Universal Crop Growth Simulator (SUCROS) (De Wit et al., 1970) and World Food Studies (WOFOST) (van Diepen et al., 1989) reveal the crop growth process and the effect of environmental factors on crop growth. These models require many crop parameters that are not readily available. Alternatively, simpler versions for describing physiological and biochemical process are used widely, such as the Decision Support System for Agrotechnology Transfer (DSSAT) model (Jones et al., 2003) and Environmental Policy Integrated Climate (EPIC) model (Williams, 1995). Due to the great applicability, these models have been used for agricultural management at a regional scale (Rosenzweig et al., 2014).

Considering the above, the objective of the study outlined in this chapter is to prove contribution of groundwater capillary rise to ET at regional scale with various crops and soils. This is important to optimize the irrigation management and design the reasonable water-saving irrigation practices to improve irrigation water productivity (IWP) by considering groundwater capillary rise to ET. The detailed objectives of this study are: (a) to develop a quantitative model requiring only readily available data that can be used to analyze the water flux at the water table, to calculate the groundwater contribution to crop evapotranspiration; (b) to indicate spatial and temporal distribution of groundwater upflow flux and effectivity of irrigation water in an irrigation district; (c) to identify the dependence of groundwater capillary rise contribution to regional evapotranspiration (F/ET) on groundwater depths; and (d) to determine the impact of irrigation amounts and groundwater depth on water productivity at regional scale.

Materials and Methods

The Study Area

Study Area and Data

The Jiefangzha Irrigation Area (JFZIA), a typical irrigation area with shallow groundwater, was chosen as the study area. With an area of 1.12 Mha, the JFZIA is a typical region of the Hetao Irrigation District, which is located in the southeast of the Lang Mountains and northwest of the Yellow River, one of the largest irrigation districts in China (Figure 6.1). The topography of this area descends from the southeast to northeast with an average slope of 0.02% (Xu et al., 2010).

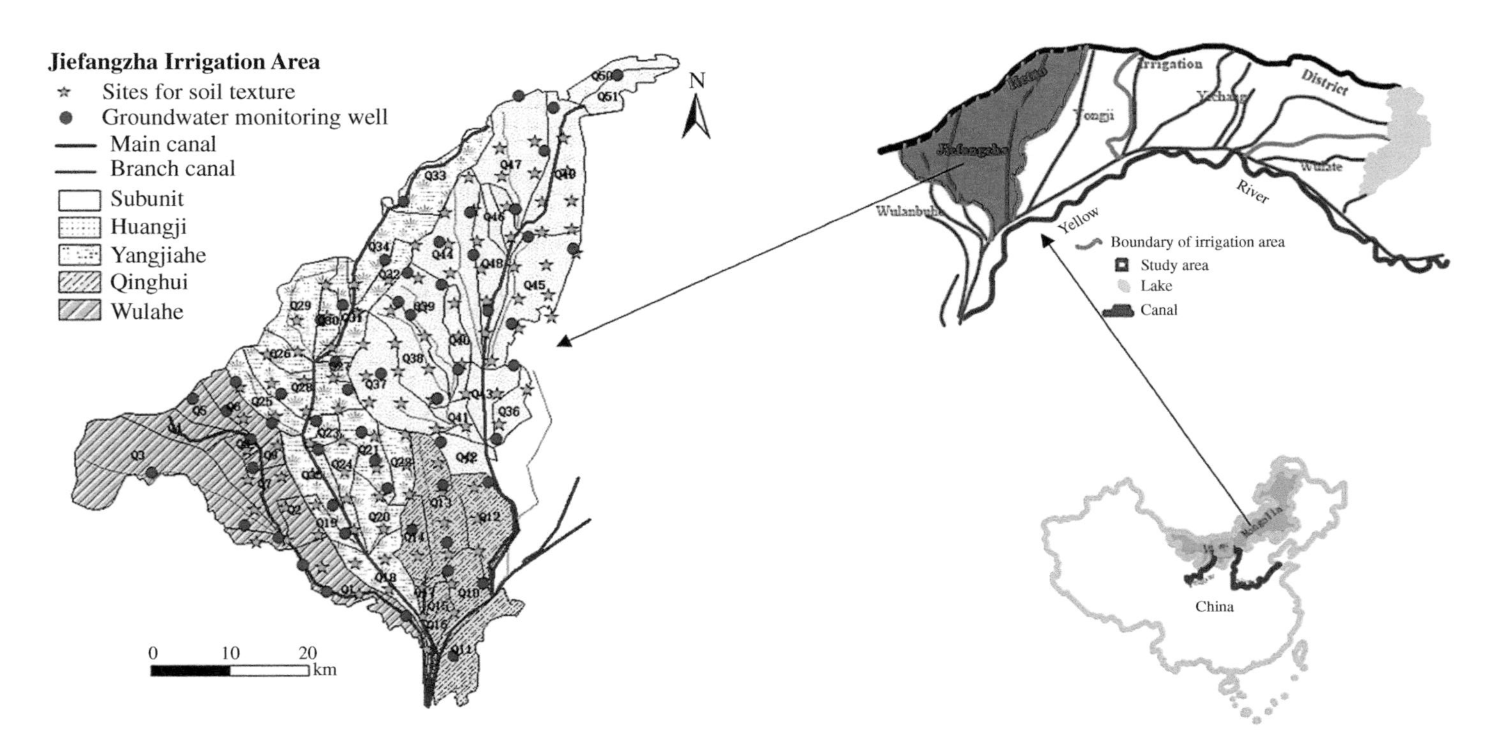

Figure 6.1 Hydrological units, location of the observation site for canal, groundwater depth, and soil texture. The dots in the left figure are sites with measured data.

The region has a typical arid and semi-arid continental climate. The region has high temperature and drought in summer along with cold and less snow in the winter. The monthly average temperature is −10.1 °C in January and 23.8 °C in July (Xu et al., 2010). The average monthly temperature from April to September ranged from 10.49 to 24.93 °C during 2009–2013. Annual average pan evaporation is approximately 2,000 mm, whereas precipitation is only 155 mm. In addition, 80% of the precipitation occurs from April to September. Irrigation is required in this area throughout the crop growing season. There are 3,100–3,300 h of sunshine and 135–150 frost-free days per year. The average elevation of this study area is 1,056 m.

Soils in the study were spatially heterogeneous with primarily silt loam and silt in the northern region and silt loam and sandy loam in the southern region. The soil begins to freeze in the middle of November and is frozen completely until late April or early May (Cai et al., 2003). In May, the soil begins to thaw, and the groundwater depth decreases.

Wheat, maize (*Zea mays* L.), and sunflower (*Helianthus annuus* L.) were the three main crops and account for 85% of total cropland in the study area. Due to low quantities of rainfall, irrigation is necessary during the crop growing season. The primary irrigation method is flood irrigation, with an average application of 8,000 m^3 ha^{-1} for the crop growing season. It was observed that there was an autumn irrigation practice with an average application of 2,600 m^3 ha^{-1} after the harvest of crops each year. The aims of the autumn irrigation are to store up water in the soil for the next crop planting. The mean soil salinity in this region was relatively low, with 3.8 ms cm^{-1} less than the mean threshold for the three crops (Allen et al., 2006), and the effect of soil salinity on crop growth was ignored in this study.

The Yellow River is the main water source of this irrigation district, with annual water supplication of approximately 47.89×10^8 m^3. The irrigation amount, irrigation time, and irrigation area were obtained from the Hetao Irrigation Administration Bureau. In recent years, water-saving irrigation was widely applied in this region. The irrigation amount varied in different regions. A total of 51 irrigation units were determined by the corresponding control canals. The distributions of the irrigation amount in the study area during 2009–2013 are shown in Figure 6.2. According to the distribution of the main canals, the study area was divided into four main sections including Wulahe, Qinghui, Yangjiahe, and Huangji. The majority of the total irrigation was focused on Wulahe and northeast of Huangji. The average irrigation amount in 2012 was lower than that in 2013, which was due to more precipitation in 2012.

The groundwater depth at 51 sites was measured every 5 d from 2009 to 2013. During the 5 yr, the groundwater depth of 51 sites ranged from 0.1 to 5.57 m in the JFZIA. The spatial distributions of groundwater depth during 2009–2013 are shown in Figure 6.2. In most of the region, the groundwater depth was shallower than 3.0 m. The average groundwater depths of the Wulahe, Yangjiahe, Huangji, and Qinghui irrigation area during 2009–2013 were 1.50, 1.98, 2.09, and 2.23 m, respectively. With the implementation of water-saving measurements, the average groundwater depth increased from 1.96 m in 2009 to 2.12 m in 2011. Then with more precipitation in 2012, the average groundwater depth became 1.78 m in 2012 and 1.92 m in 2013.

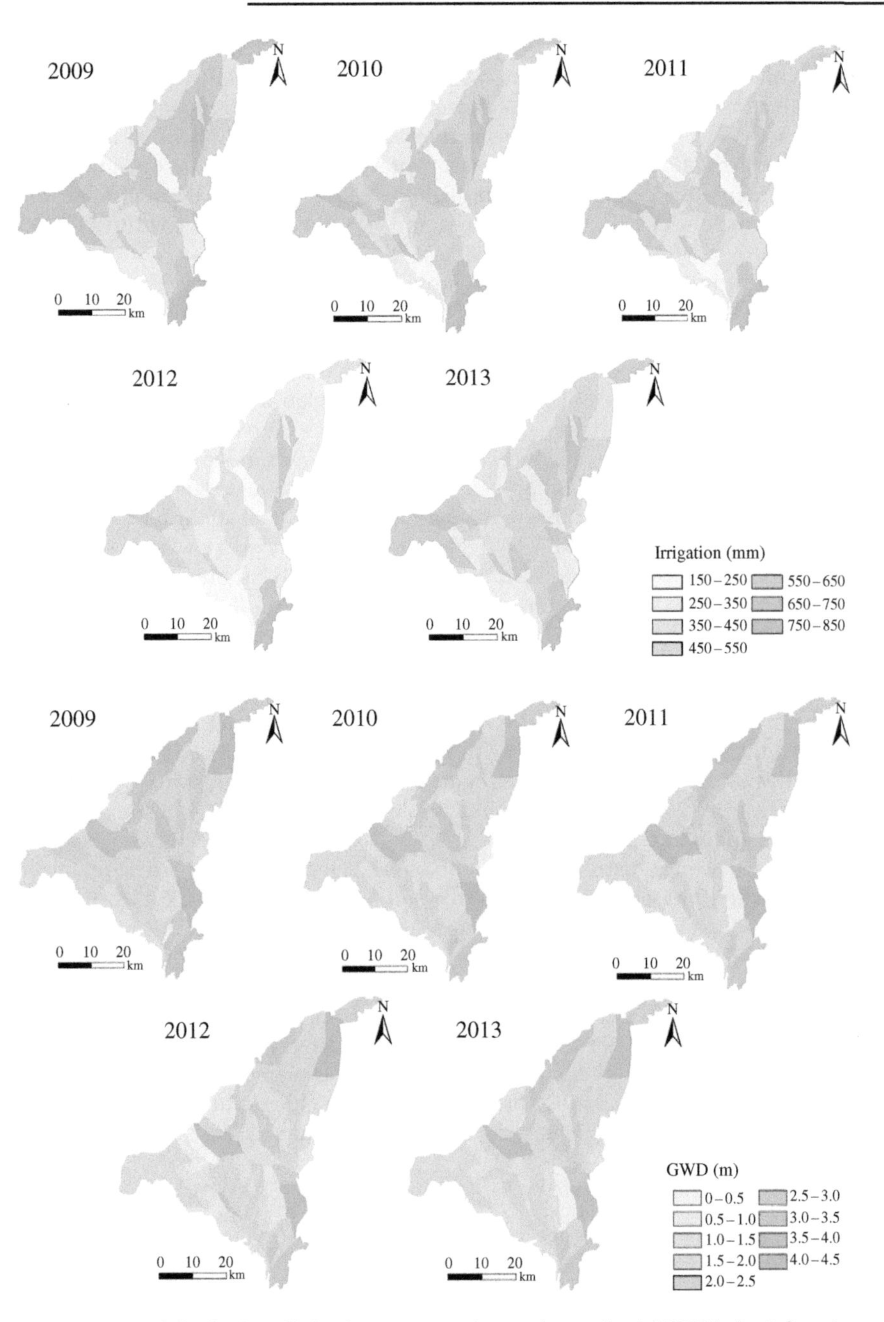

Figure 6.2 Spatial distribution of irrigation amount and groundwater depth (GWD) in the Jiefangzha Irrigation Area during 2009–2013.

Experiments and Data

For model calibration and validation, field experiments by Kong (2009) were used. Investigating the effect of water table depth (ranging from 1.5 to 3 m) on water use efficiency (WUE) of maize using lysimeters at the Shuguang Experimental farm in the Hetao Irrigation District. The experimental station is located at open field and has representative climate, soil, and groundwater conditions. Lysimetric experiments provide accurate water balances and have been widely used to validate models.

The experiment with eight nonweighing lysimeters grown with maize were carried by Fanrui Kong in 2007 and 2008 in the middle of the experimental field with groundwater depth initially at 1.5, 2, 2.5, and 3 m replicated two times (Kong, 2009). The iron cylindrical lysimeters had a diameter of 1 m and a height of 3.35 m. The maize (Kehe-8) commonly grown in Hetao was planted in late April and harvested in late September. Cultivation practices were similar to the practices by farmers and recommended by extension agents. Four times during the growing season 360 mm of water was applied. The irrigation amount referred to the local farmers' field. Nitrogen fertilizers were applied at a rate of 207 kg ha^{-1} at the first irrigation, 103 kg ha^{-1} at the third irrigation in 2007, and 276 kg ha^{-1} at the first irrigation in 2008. Marriotte bottles were used to set the initial groundwater depths at 1.5, 2, 2.5, and 3 m before sowing, and then removed to allow the groundwater to change in response to irrigation and evaporation.

During the growing period, the meteorological data consisting of air temperature, sunshine hours, relative humidity and wind speed for 2007 and 2008 were taken from Linhe weather station 8 km away from Shuguang and were used to calculate the FAO Penman–Monteith reference evapotranspiration (ET_0). Daily rainfall was measured using a rain gauge at the experimental site. The rainfall was 132 mm in 2007 and 123 mm in 2008, with more than 70% of precipitation occurring in July and August.

The particle-size distribution was obtained using the laser particle analyzer. The main soil physical properties of the soil are shown in Table 6.1. In the root zone, the soil is loam, which is the representative soil texture in this region (Ren et al., 2016). The dry bulk density was obtained by oven drying undisturbed soil samples of 100 cm^3 at 105 °C for 48 h. Saturated hydraulic conductivity was measured in eight 100 cm^3 undisturbed soil samples using a constant-head permeameter (Wit, 1967). Soil water retention curves were determined for each horizon in 100 cm^3 undisturbed soil samples using a pressure membrane apparatus (SEC-1000). The fitted soil hydraulic parameters of the Muchlem–van Genuchten model are presented in Table 6.2 (Mualem, 1976; van Genuchten, 1980).

Table 6.1 Soil physical properties of experimental area

	Soil particle-size distribution						
Soil depths cm	Clay (<0.002 mm)	Silt (0.002–0.05 mm)	Sand (0.05–2.0 mm)	Soil texture	Bulk density g cm^{-3}	Field capacity, *mf* cm^3 cm^{-3}	Wilting point
0–90	17.8	29.4	52.8	Loam	1.27	.32	.04
90–150	8.1	39.1	52.8	Sandy loam	1.32	.30	.06

Table 6.2 Values of the hydrological parameters for AWPM-SG

Depths cm	Layer and soil type	*ms* cm^3	*md* cm^{-3}	*ks* cm d^{-1}	*C*	dp	D_0	*b*
Initial values								
0–90	Loam	0.395	.07	15	8		.88	35.4
90–300	Sandy loam	0.43	.07	14		.08		
Calibrated values								
0–90	Loam	0.395	.02	15	13		.1	17.5
90–300	Sandy loam	0.41	.02	23		.2		

Note: AWPM-SG, Agricultural Water Productivity Model for Shallow Groundwater; *ms* and *md* are the saturated moisture and residual moisture of 0–90 cm soil, respectively; *ks* is the saturated hydraulic conductivity; *C* and *b* are the content of soil in Zones 1 and 2; D_0 is the diffusion rate at wilting point; dp is the drainable porosity.

The soil moisture content was measured every 3–17 d using the time domain reflectometry (Tube-TDR) including shortly before and after the irrigation and at seeding and harvest. The soil moisture was measured at 20-cm intervals from the surface to a depth of 300 cm. At the same time, it was calibrated with drying method through earth-fetching at regular time.

The groundwater depth was monitored daily during the crop growth period through monitoring the piezometric head in the access chamber.

The crop leaf area index (LAI) was measured every 6–12 d using a leaf area meter (LI-3000, LI-COR).

Dry maize yield was determined after harvesting.

Method

Agricultural Water Productivity with Shallow Groundwater Model at Field Scale

In this study, we develop the Agricultural Water Productivity Model for Shallow Groundwater (AWPM-SG), which is coupling of a crop growth model (EPIC) and a soil moisture model of root and vadose zone (WIPE, Watershed Irrigation Potential Estimation) that simulates both upward movement from groundwater and percolation to the groundwater, which was originally developed by Saleh et al. (1989) for surface irrigation in Bangladesh (Williams, 1995). The structure of the AWPM-SG model is shown in Figure 6.3. The AWPM-SG model needs few parameters and simulates water flux on daily time step. The three parts of the AWPM-SG model consists of a crop module, an actual evapotranspiration module, and a soil module. These three modules are coupled first time in the AWPM-SG model. An overview of the AWPM-SG model is given below.

Crop Module

The crop module of AWPM-SG is mainly based on the EPIC model, which was originally developed by Williams et al. (Williams et al., 1989). It has been tested and applied widely around the world (Rosenberg et al., 1992; Mearns et al., 1999; Carbone et al., 2003; Easterling et al., 1993; Liu et al., 2007). Wang and Li reported that the EPIC model predicted winter wheat and spring maize yield well on the Loess Plateau in China (Wang & Li, 2010). Niu et al. (2009) examined the reliability

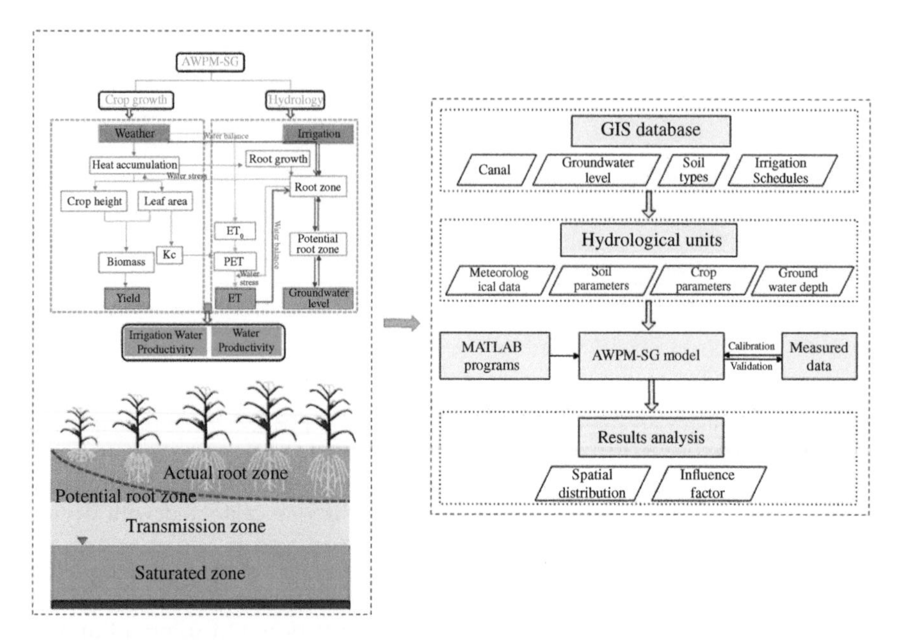

Figure 6.3 The schematic of regional simulation using the Agricultural Water Productivity Model for Shallow Groundwater (AWPM-SG) model.

of the EPIC model in simulating the grain sorghum [*Sorghum bicolor* (L.) Moench] yields in the USA and found that the accuracy and reliability varied with climate classes and nitrogen treatments. Brown and Rosenberg (1999) validated the EPIC model to assess the climate change impact on the potential productivity of corn and wheat in the US. Currently the EPIC model has evolved into a comprehensive model capable of simulating photosynthesis, evapotranspiration, and other major plant and soil process (Wang & Li, 2010). The crop module includes the phonological development, crop growth indexes (LAI, biomass, root growing, crop yield), and water productivity (WP, IWP).

Description of Soil Module

The EPIC model was coupled with the modified WIPE model to simulate the flux of the groundwater–soil water–crop water consumption system. The WIPE model was developed by Saleh et al. (1989) to study the impact of irrigation management schemes on groundwater levels in Bangladesh. Precipitation, irrigation, soil properties such as moisture content and hydraulic conductivity, and the initial groundwater level are required to run this model. In this mode, the soil profile was divided into four zones: the actual root zone (Zone 1), potential root zone (Zone 2), transmission zone (Zone 3), and the saturated zone (Zone 4). Zone 1 is the zone occupied by roots; Zone 2 is the zone that is not currently occupied by the roots but will be so after their complete development; Zone 3 is the unsaturated transition zone below the root zone with a lower boundary at water table, and the thickness of this layer varies in time according to extraction/evaporation and recharge; and Zone 4 is the saturated zone and is regarded as the water table. Zone 3 is always at a constant moisture content and is equal to the saturated moisture content minus the drainable porosity. During the simulation process, the soil texture in Zones 1 and 2 were the same. The water balance method was used for the water movement calculation of Zone 1.

In this study, measured groundwater depths were used as input. The soil module was a one-dimensional model employing the Thornthwaite–Mather procedure to calculate the recharge below the root zone to the aquifer, which is primarily applicable to shallow aquifers (Steenhuis & Van Der Molen, 1986).

Actual Evapotranspiration Module

Actual evapotranspiration is an input to the EPIC model. We used the model of Kendy et al. (2003), which previously was able to calculate the actual evapotranspiration from soil–water storage in the North China Plain with good accuracy. In this model, the ratio of E_p to T_p depends upon the development stage of the leaf canopy, moisture content, and root development (Sau et al., 2004; Kendy et al., 2003; Campbell & Norman, 1998). The input data are the daily LAI, LAI simulated by the EPIC model, and simulated soil moisture of actual root zone (Zone 1) by the WIPE model and water balance.

Division of Hydrological Response Unit at Regional Scale

Due to the low spatial heterogeneity of meteorological data in the JFZIA with flat topography, the same meteorological data were used for the entire study area. Considering the distributions of branch canals, sub-lateral canals, groundwater depth, and soil texture, the study area was divided into 51 hydrological response units (HRUs) (Wulahe, Q1-Q9; Qinghui, Q10-Q17; Yangjiahe, Q18-Q35; and Huangji, Q36-Q51) (Figure 6.1).

Due to the lack of an actual land use data, the wheat, maize, sunflower, and uncultivated areas were simulated with proportions of 23, 38, 33, and 6%, respectively, for each HRU (Sun, 2014). The irrigation water depth of each HRU was obtained from canal water dividing the control area. The spatial distribution of HRUs is shown in Figure 6.1.

The model was run for each HRU with measured groundwater levels from 2009 to 2013. After the simulations, the groundwater upward flux, crop ET, crop yield, and WP for each HRU were calculated as weighted averages according to the ratios of wheat, maize, sunflower, and uncultivated land. The spatial distribution of hydrological elements during 2009–2013 was analyzed using element conversion in ArcGIS 10. The structure of the AWPM-SG model at regional scale is shown in Figure 6.3. Then the relationships between the water use indicators and the groundwater fluxes or irrigation amount were investigated using results from 2009 to 2013.

Model Calibration and Validation

Model Calibration and Validation at Field Scale

The simulation period for maize was from late April to late September in 2007 and 2008 using the observed initial soil water content and groundwater depth subject to

Table 6.3 Default and calibrated values of maize's physiological parameters for the crop growth part of AWPM-SG

Parameters	Default values	Calibrated values
dimensionless canopy extinction coefficient, *kb*	.5	.2
b_e	.3	.8
b_t	4	3
Minimum temperature for plant growth, T_b, °C	8	8
Optimal temperature for plant growth, T_0, °C	25	25
Leaf area index decline rate, *ad*	1.0	0.5
Maximum crop height, h_{mx}, cm	200	225
Maximum leaf area index, LAI_{mx}	6.0	5.5
Maximum root depth, RD_{mx}, cm	90	90
Plant radiation-use efficiency, BE, (kg ha^{-1}) (MJ m^{-2})$^{-1}$	40	40
Harvest index, HI	.5	.5
Total potential heat units required for crop maturation, PHU, °C	2,000	2,100
A parameter expressing the sensitivity of harvest index to drought, WYSF	.05	.05

Note. AWPM-SG, Agricultural Water Productivity Model for Shallow Groundwater.

the imposed irrigation schedule and fertilizer applications for 2 yr. The 2007 data were used for calibration and the 2008 data were used to validate the model. Soil moisture of the top 90 cm (Zone 1 and 2), groundwater depth, and LAI were simulated. The soil hydraulic parameters (*md, ms, C*, α, and *ks*) and the crop parameters were calibrated (Tables 6.2 and 6.3). During the calibration, we set the parameters of model according to the measured data and recommended values, and we analyzed the sensitivity and uncertainty analysis of parameters and found the sensitive parameters for soil water, groundwater depth, and crop LAI, respectively, such as LAI_{mx}, *mf*, and so on. Then we adjust the parameters as their sensitivity to make the simulation result of the model closer to the measured data. At last the calibrated parameters were used to validate the model using the data of 2008. The default values of the EPIC model for maize were used as initial value for simulating crop growth (Williams et al., 2006). The default value of maximum rooting depth was 90 cm measured in the experiment by Kong (Kong, 2009). Initial soil water content and groundwater depth were specified according to measurements. A sensitivity and uncertainty analysis of parameters was performed for soil water content, groundwater depth, and crop LAI.

The upper boundary condition was determined by the actual evaporation and transpiration rates, and the irrigation and precipitation fluxed. A no-flux boundary condition was specified at the column bottom.

The mean relative error (MRE), the root mean square error (RMSE), the Nash and Sutcliffe model efficiency (NSE), the coefficient of determination (R^2), and the coefficient of regression (b) were used to quantify the model-fitting performance for both calibration and validation processes. These indicators were defined as follows (Raes et al., 2012; Xu et al., 2015):

MRE is the mean relative error. The MRE close to 0 indicates good model predictions.

$$\mathrm{MRE} = \frac{1}{N}\sum_{i=1}^{N}\frac{(P_i - O_i)}{O_i} \times 100 \quad (6.1)$$

RMSE is the root mean square error. The RMSE value close to 0 indicates good model predictions.

$$\text{RMSE} = \sqrt{\frac{1}{N}\sum_{i=1}^{N}(P_i - O_i)^2} \tag{6.2}$$

NSE is the Nash and Sutcliffe model efficiency. When NSE is 1.0, it represents a perfect fit, NSE close to 0 represents the predicted values near to the averaged measurement, and negative NSE values indicate that the mean observed value is a better predictor than the simulated value

$$\text{NSE} = 1 - \frac{\sum_{i=1}^{N}(P_i - O_i)^2}{\sum_{i=1}^{N}(O_i - \bar{O})^2} \tag{6.3}$$

where R^2 is the coefficient of determination; R^2 values close to 1 indicate good model predictions.

$$R^2 - \left[\frac{\sum_{i=1}^{N}(O_i - \bar{O})(P_i - \bar{P})}{\left[\sum_{I=1}^{N}(O_i - \bar{O})^2\right]^{0.5}\left[\sum_{I=1}^{N}(P_i - \bar{P})^2\right]^{0.5}}\right]^2 \tag{6.4}$$

where b is the coefficient of regression; b values close to 1 indicate good model predictions

$$b = \frac{\sum_{i=1}^{N} O_i \times P}{\sum_{i=1}^{N} O_i^2} \tag{6.5}$$

where N is the total number of observations, P_i and O_i are, respectively, the ith predicted and observed values ($i = 1, 2, \ldots, N$), and $\bar{P}$ and $\bar{O}$ are the predicted and observed mean values, respectively (Moriasi et al., 2007).

Sensitivity Analysis of Parameters

Through increasing or decreasing of the percentage of the parameters, the variation of ET, groundwater depth, LAI, and soil water of 90 cm are analyzed. Then the sensitivity of each parameter on ET, groundwater depth, LAI, and soil water of 90 cm can be obtained.

Uncertainty Analysis of Parameters

The d-factor was used to analyze the uncertainty of parameters (Abbaspour et al., 2004; Talebizadeh & Moridnejad, 2011). The d-factor is indicative of average distance between the upper and lower confidence interval (in this study the 95% prediction interval). The d-factor was calculated as follows:

$$\bar{d}_x = \frac{1}{n}\sum_{i=1}^{n}(X_{Ui} - X_{Li}) \tag{6.6}$$

$$d\text{-factor} = \frac{\bar{d}_x}{\sigma_x} \tag{6.7}$$

where $\bar{d}_x$ is the average distance between the lower X_L and the upper limits X_U of the confidence interval, σ_x is the standard deviation of observed data, and n is the number of data. Larger d-factors will lead to larger uncertainty.

Model Calibration and Validation at Regional Scale

Based on the model calibration and validation at field scale, measurements of soil moisture content for 18 sites each year, LAI for 2 sites each year, and yield for 8 sites in 2012 and 2 sites in 2013 over the crop growing period of the study area were used to calibrate and validate the AWPM-SG model at the regional scale (Figure 6.1). The 2012 data were used for calibration and 2013 data were used for validation of the model. The model reproduces the soil moisture of root zone (Zones 1 and 2), LAI, groundwater upward flux, ET, and crop yield using the observed initial soil moisture content and groundwater depth subject to the irrigation schedules and precipitation. Soil hydraulic parameters (residual soil moisture, θ_r; saturated soil moisture, θ_s; field capacity, θ_{fc}; soil moisture at wilting point, θ_{wp}; soil constant, C and α; and saturated hydraulic conductivity, ks) and crop parameters (maximum leaf area index, LAI_{mx}; extinction coefficient of the canopy, k_b; empirical parameters for evaporation and transpiration, b_e and b_t; energy conversion factor, BE; and harvest index, HI) were calibrated. The initial crop parameters for simulating crop growth were used as the default values in the EPIC model for wheat, maize, and sunflower. Initial soil moisture content and groundwater depth were specified according to measurements.

Water Productivity and Irrigation Water Productivity

Water productivity (WP) is the ratio of crop yield to crop ET during the crop growing season, which was calculated as:

$$\text{WP} = \frac{Y}{\text{ET}} \tag{6.8}$$

where Y is the crop yield (kg ha^{-1}), and ET is the total crop evapotranspiration during the crop growth period (mm).

Irrigation water productivity (IWP) is defined as the ratio of crop yield (Y) to the seasonal application of irrigation water (I) (Salah et al., 2014), calculated as:

$$\text{IWP} = \frac{Y}{I} \tag{6.9}$$

Results

Evaluation of the AWPM-SG Model

Evaluation of the AWPM-SG Model at Field Scale

Data of a lysimetric experimental study carried in 2007 and 2008 at the Shuguang experimental station in the Hetao irrigation district, were used to calibrate and validate the AWPM-SG model. The maize (Kehe-8) commonly grown in Hetao irrigation district was planted in late April and harvested in late September. Four times irrigation with 97.50, 90, 97.50, and 75 mm were applied in 2007 and 2008, respectively. Experiments were carried out in two replicates. The groundwater depths were set using the Marriotte bottles at 1.5 , 2, 2.5, and 3 m before planting. After planting the bottles were removed, allowing the groundwater to vary in response to irrigation and evaporation.

The 2007 data were used for calibration and 2008 data for validation (Tables 6.2 and 6.3). The soil data and crop data inputted are shown in Tables 6.2 and 6.3. The goodness of fit test can be found in Table 6.4. In cases where the observed data remained nearly the same during the growing season, such as the deeper groundwater depth, the Nash Sutcliff efficiency gives unrealistic results.

Soil Moisture: Calibration and Validation

The soil water content of the top 90 cm (Zones 1 and 2) was used to calibrate and validate the model. Simulated mean soil water content in 2007 in the 90-cm soil zone for groundwater depth at 150 and 200 cm were generally satisfactory simulated with the NSE of .57 and .68, and R^2 of .57 and .69 (Figure 6.4A[a], [b]). For deeper water table depths of 250 and 300 cm, the simulated water content fitted visually well (Figure 6.4A[c], [d]) and the goodness of fit was reasonable for most statistics presented in Table 6.4. In 2008 the soil moisture was well predicted with the calibrated input data (Figure 6.4A[e], [h], Table 6.4) including the NSE, which varied from .39 to .74 and R^2 varying from .77 to .88 (Table 6.4). For the calibration and validation, the averaged RMSE value was 2.97 cm, varying from 2.09 to 4.16 cm.

Groundwater Table: Calibration and Validation

Figure 6.4B shows that the AWPM-SG simulations of the groundwater table followed the same trend with observations during the calibration period as indicated by the goodness of fit parameters with the R^2 around .6 and averaged RMSE value of 9.6 cm (Table 6.4). The calibration period the simulated data in 2008 (Figure 6.4B) had a similar accuracy with the MRE varying from −8.28 to 3.28%. (Table 6.4).

Crop Leaf Area Index: Calibration and Validation

Figure 6.4C[a–d] shows that the simulated LAI followed the same trend as the observations during the calibration period. The NSE for LAI varied from .88 to .97 and R^2 varied from .88 to .98, indicating a good fit (Table 6.4). The averaged RMSE value for calibration and validation was 0.6 cm^2 cm^{-2}. The measured LAI in 2008 was less than in 2007, which can have many causes that were not included in the model, such as the low temperature and snow in the seeding stage as well as different observers in 2007 and 2008. The LAI was therefore not as well predicted by AWPM-SG for the validation period as for the calibration period with the NSE ranging from .18 to .88 (Figure 6.4C[e–h]).

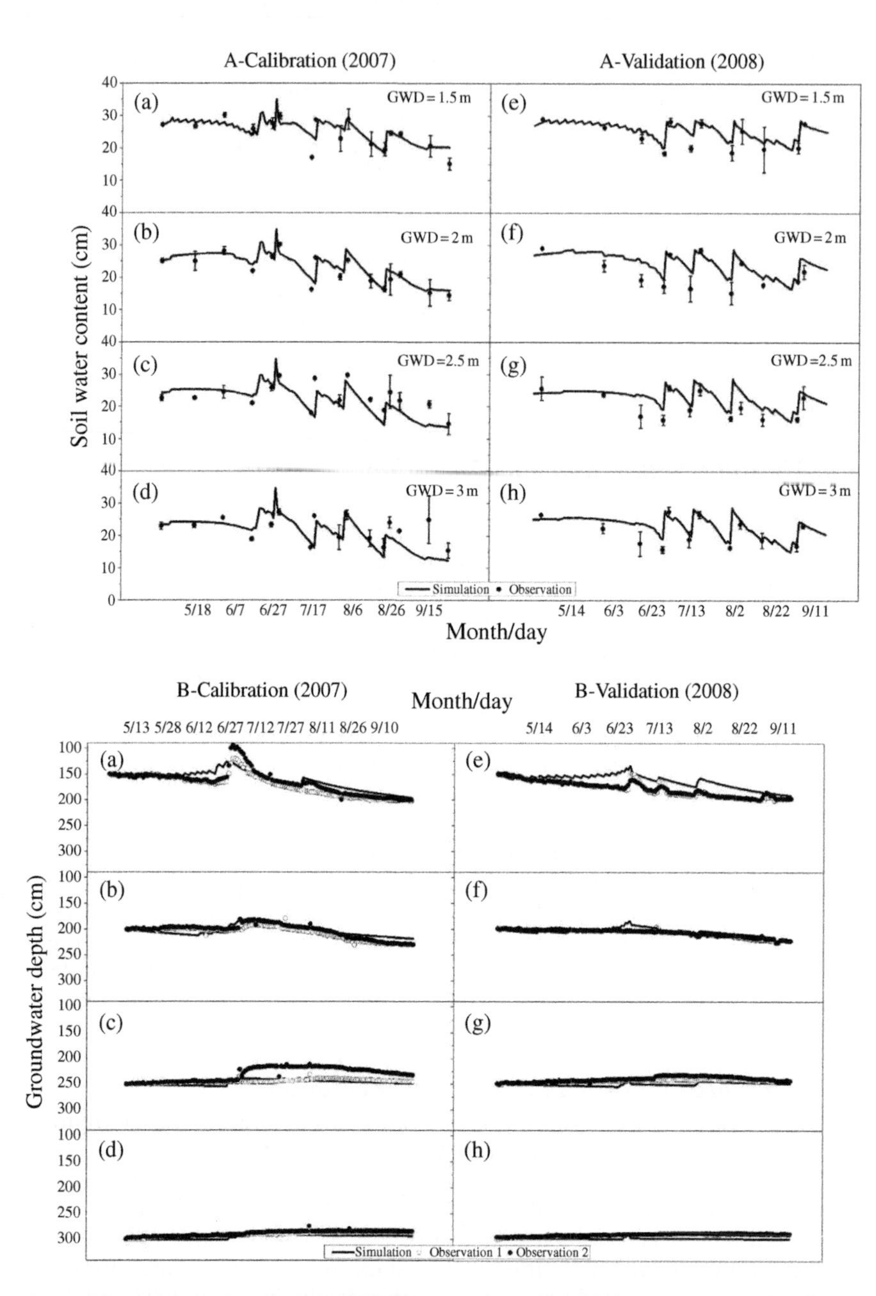

Figure 6.4 (A, B, C) Simulated vs. measured soil water content, groundwater depth (GWD), and leaf area index under different groundwater depths during calibration (a–d) in 2007 and validation (e–h) in 2008. A is that simulated vs. measured soil water content, B is that simulated vs. measured groundwater depth, and C is that simulated vs. measured leaf area index.

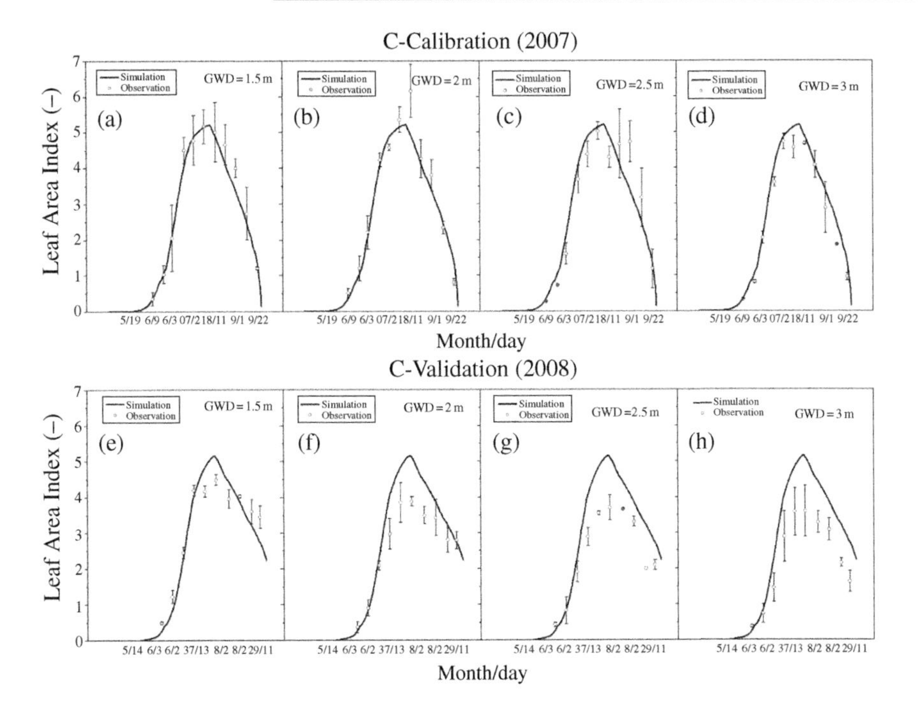

Figure 6.4 ***(Continued)***

Parameter Sensitivity and Uncertainty Analysis

To investigate the effect of calibrated parameters on the crop evapotranspiration, groundwater depth, LAI, and soil water content of 90 cm soil, the parameters sensitivity and uncertainty were analyzed together. When the parameters vary, the four indexes are linearly related to the change in the parameters basically except for the temperature for plant growth, T_b; potential heat units, PHU; and C. For the crop evapotranspiration, the LAI_{mx} affects the ET most obviously, and then the b_e and T_b largely affect the ET (Figure 6.5a,b). The other parameters affect the ET slightly with the $\Delta ET/ET$ less than 5% as the parameters increase by 25%. Most of the parameters have little impact on the ET. The initial LAI_{mx} is from the recommended value of EPIC. Through the calibration and sensitivity of parameters the final LAI is determined.

For the groundwater depth, the *mf*; drainable porosity, dp; T_b; and PHU are the main impact factors. The others almost have no influence on the groundwater depth (Figure 6.5c,d), and mf is measured in the field experiment using undisturbed samples. The initial dp is obtained by the recommended value of local government and T_b, PHU is obtained by the recommended value of EPIC.

Similar to the ET, the main impact factors on LAI are LAI_{mx}, T_b, PHU, T_0, and LAI decline rate, ad. The other parameters have little effect on the LAI. The initial T_b, PHU, T_0, and ad are recommended by EPIC.

Table 6.4 Mean relative error, root mean square error, regression coefficient, Nash and Sutcliffe model efficiency, and coefficient of determination of the model

Variable	GWD	Calibration, 2007			Validation, 2008		
		Soil water content	Groundwater depth	LAI	Soil water content	Groundwater depth	LAI
	–cm–						
Mean relative error, MRE, %	150	1.77	−2.13	−2.72	6.2	−8.28	−7.68
	200	2.67	1.46	−0.81	10.8	−0.52	12.63
	250	−8.38	4.11	8.75	12.5	3.21	22.78
	300	−7.05	1.92	15.04	5	2.58	33.89
Root mean square error, RMSE (cm cm^{-2} cm^{-2})	150	2.93	12.39	0.3	2.09	17.5	0.45
	200	2.7	8.97	0.46	3.03	5.29	0.7
	250	4.16	10.9	0.59	3.05	8.57	0.89
	300	3.7	6	0.36	2.11	8.05	1.01
Regression coefficient, *b*	150	0.98	0.97	0.95	1.04	0.91	1.05
	200	1	1.01	0.93	1.07	0.99	1.21
	250	0.91	1.04	0.98	1.1	1.03	1.3
	300	0.93	1.02	1.08	1.02	1.03	1.4
Nash and Sutcliffe model efficiency, NSE	150	0.57	0.67	0.97	0.72	−0.53	0.88
	200	0.68	0.58	0.94	0.57	0.44	0.62
	250	−0.01	−0.66	0.88	0.39	−4.42	0.37
	300	0	−0.97	0.95	0.74	−7.64	0.18
Coefficient of determination, R^2	150	.57	.72	.98	.88	.61	.93
	200	.69	.64	.95	.77	.56	.96
	250	.44	.4	.88	.79	.14	.96
	300	.51	.66	.98	.79	.01	.97

Note: GWD, groundwater depth; LAI, leaf area index.

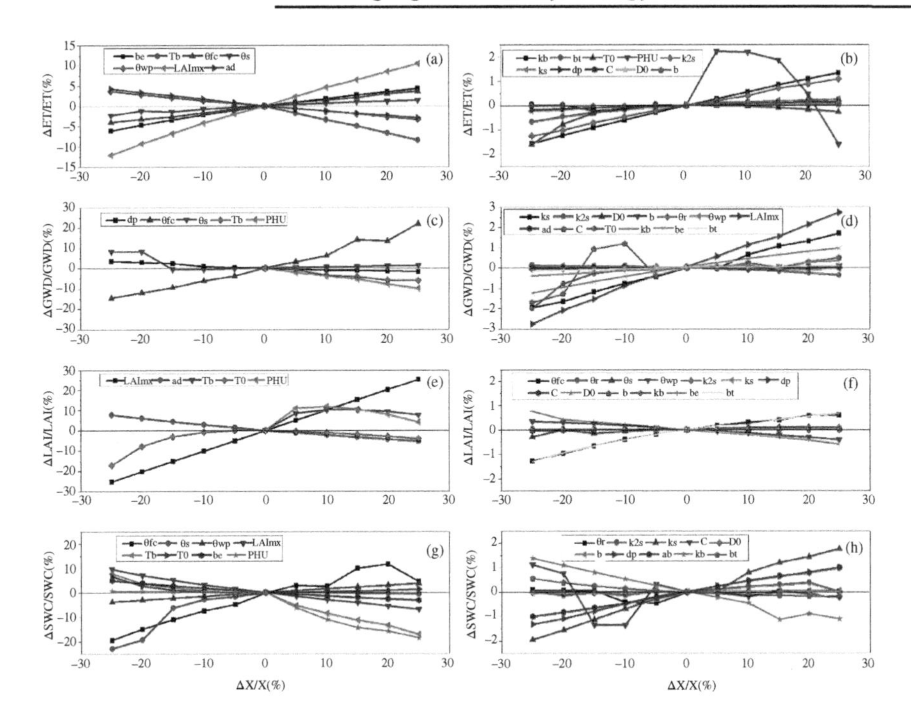

Figure 6.5 Parameters sensitivity analysis for evapotranspiration (ET), groundwater depth, leaf area index (LAI), and soil water content.

For the soil water content of 90 cm, the parameters such as *mf*, *ms*, wilting point of soil moisture (mwp), LAI_{mx}, PHU, and T_b affect the soil water content more obviously than the other parameters; *mf*, *ms*, and mwp are measured in the experiment. The initial LAI_{mx}, PHU, and T_b are recommended by EPIC.

Combing the observed data, the uncertainty analysis of the parameters on groundwater depth, LAI, and soil water content are analyzed. Through the uncertainty analysis of parameters, the *d*-factors calculated by Equation 6.7 ranges from 0 to 1.44 with an average value of 0.14. A larger *d*-factor represents larger uncertainty. From the data, we can find that for the groundwater depth, LAI, and soil water content, the most uncertain parameters are the same with the most sensitive parameter. The most uncertain parameters are *mf*, LAI_{mx}, *mf* for groundwater depth, LAI, and soil water content, respectively. The trend of uncertainty of parameters is basically similar to the sensitivity.

Evaluation of the AWPM-SG Model at Regional Scale

The data in 2012 were used for calibration and data in 2013 for validation of the AWPM-SG model. Performance of the model can be found in Figure 6.6. The results showed that the simulated and measured data were distributed evenly on both sides of the 1:1 line with NSE values of .72, .90, .7, and .69 for soil water

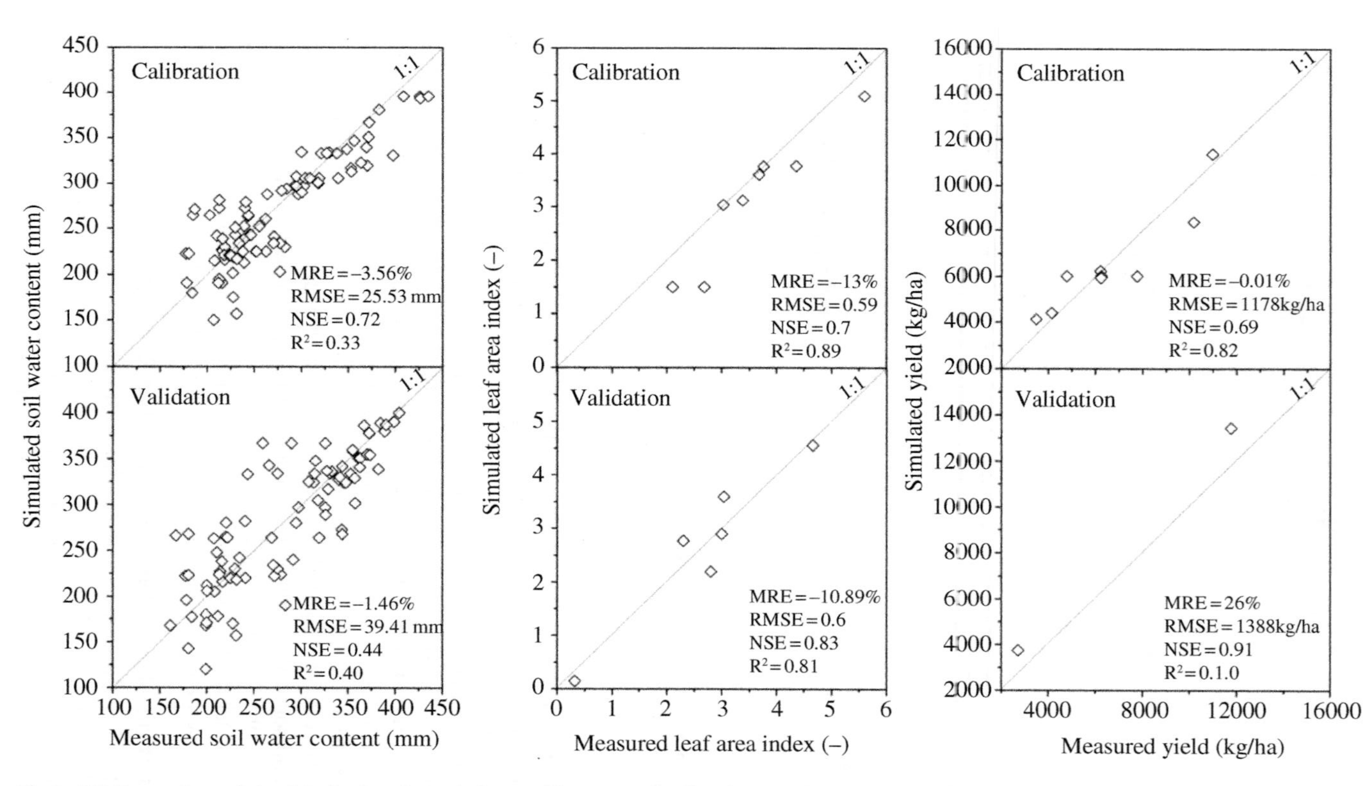

Figure 6.6 Comparison of simulated soil and crop indexes with measured soil and crop indexes in 2012 and 2013.

content, groundwater depth, crop LAI, and crop yield, respectively. In Xu's study (Xu et al., 2010), the simulated soil water content, groundwater depth, and LAI showed agreement with the measured values, resulting in NSE values of .61, .81, and .99, respectively. In addition, the RMSEs were 25.53 mm soil water content, 0.43 m groundwater depth, 0.59 crop LAI, and 1,178 kg ha^{-1} crop yield. Therefore, we considered that the simulation results were accurate enough for our regional study.

To test the reasonability of the soil parameters and crop parameters, the soil water content, groundwater depth, crop LAI, and crop yield at 18 sites in 2013 were used for validation (Figure 6.6). The results showed that the MRE for soil water content, groundwater depth, crop LAI, and crop yield ranged from −10.89 to 26%. The R^2 for the four indicators were all greater than .4, and NSE values were all greater than .44. Due to minimal yield data in 2013, the error for crop yield in the validation process was slightly greater.

Groundwater Contribution to ET

The spatial distributions of field ET during 2009–2013 are shown in Figure 6.7a. Evapotranspiration ranged from 495.61 to 787.47 mm during the period of April to September for the 5 yr. At the regional scale, the crop ET values in the central parts of the study area were higher than those in the other parts. In the calculation of this model, groundwater upward flux was related to soil saturated hydraulic conductivity positively, then low soil saturated conductivity led to less groundwater upward flux and crop ET in the western part of the study area. In the eastern part of the study area deep groundwater led to less capillary rise, which resulted in less crop ET (Figure 6.2).

Shallow groundwater contribution to crop growth is important to determine irrigation scheduling. Here, groundwater upward flux was used as the contribution of groundwater to ET. In this study, the groundwater upward flux (*F*) was the net upward flux at the water table, which is the upward flux minus the downward flux at the water table. The spatial distributions of groundwater upward flux (net groundwater contribution to soil water, *F*) during 2009–2013 are shown in Figure 6.7b. At the regional scale, groundwater upward flux gradually decreased from west to east. The maximum groundwater upward flux was 248.04, 298.23, 300.10, 90.13, and 172.74 mm during 2009–2013, respectively. Because of more rainfall in 2012, the groundwater upward flux was less than those in other years. Then, in the Hetao irrigation district with shallow groundwater, the groundwater contribution to crop growth was significant.

Furthermore, groundwater upward flux has a significant seasonal trend from 2009 to 2013 (Figure 6.8). In May and June, the groundwater upward flux was negative, meaning the percolation of soil water was greater than groundwater upward flux. This was because in May and June, the crops have enough irrigation water supplied compared with the field crop water use, whereas in July and August, the crop ET was greater than irrigation, so the groundwater upward flux at this time was much greater than that in the other period. In September, the groundwater upward flux and crop ET began to decrease.

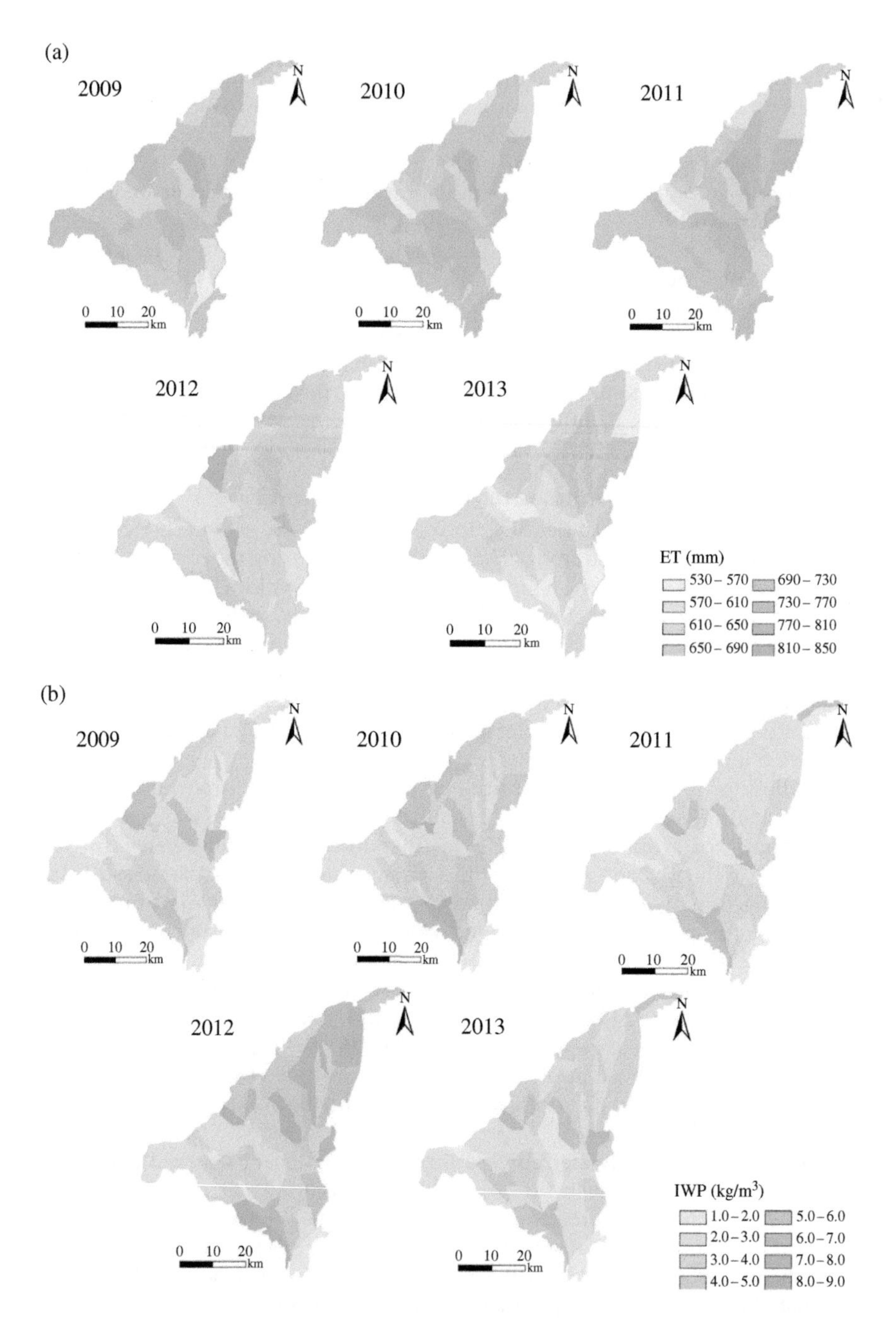

Figure 6.7 Spatial distribution of (a) crop evapotranspiration and (b) groundwater upward flux in the Jiefangzha Irrigation Area (JFZIA) during 2009–2013.

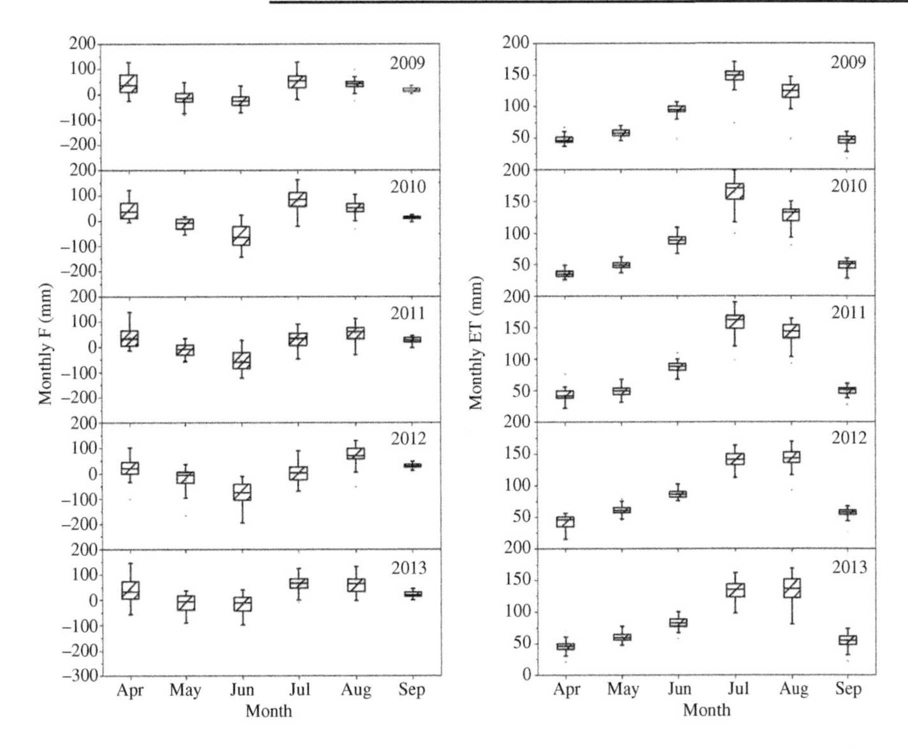

Figure 6.8 Temporal variation of monthly F and ET during 2009–2013. F, net groundwater contribution to soil water, which is the capillary rise water minus downward flux at water table ; "+" refers to that the groundwater capillary rise to soil water is more than the soil water percolating to groundwater; "–" refers to that the groundwater capillary rise to soil water is less than the soil water percolating to groundwater; ET, crop evapotranspiration.

Discussion

Relationship between Groundwater Upward Flux to Evapotranspiration and Average Groundwater Depth

To understand the seasonal trend of the impact of groundwater depth on F/ET, the relationship between the monthly groundwater contribution to crop ET (F/ET) and groundwater depth from 2009 to 2013 was analyzed in Figure 6.9a. The results indicated that the F/ET decreased from 45% in April to –80% in June and increased from –80% in June to 50% in September of 2012. However, groundwater became shallower from April to June due to the recharge from soil water thawing. A significant trend was found where groundwater depth declined from 1.3 m in June to 1.8 m in September of 2012, and this could be attributed to the groundwater upward flux due to crop evapotranspiration. Although groundwater became deeper and deeper, a higher water requirement of the crops resulted in the groundwater contribution to crop water consumption (F/ET) increasing from June to September. Additionally, crop development stage is an important factor affecting groundwater evaporation (Wang et al., 2016). The high F/ET values of 87, 97, 85, 40, and 75% in April from 2009 to 2013, respectively, were due to the groundwater

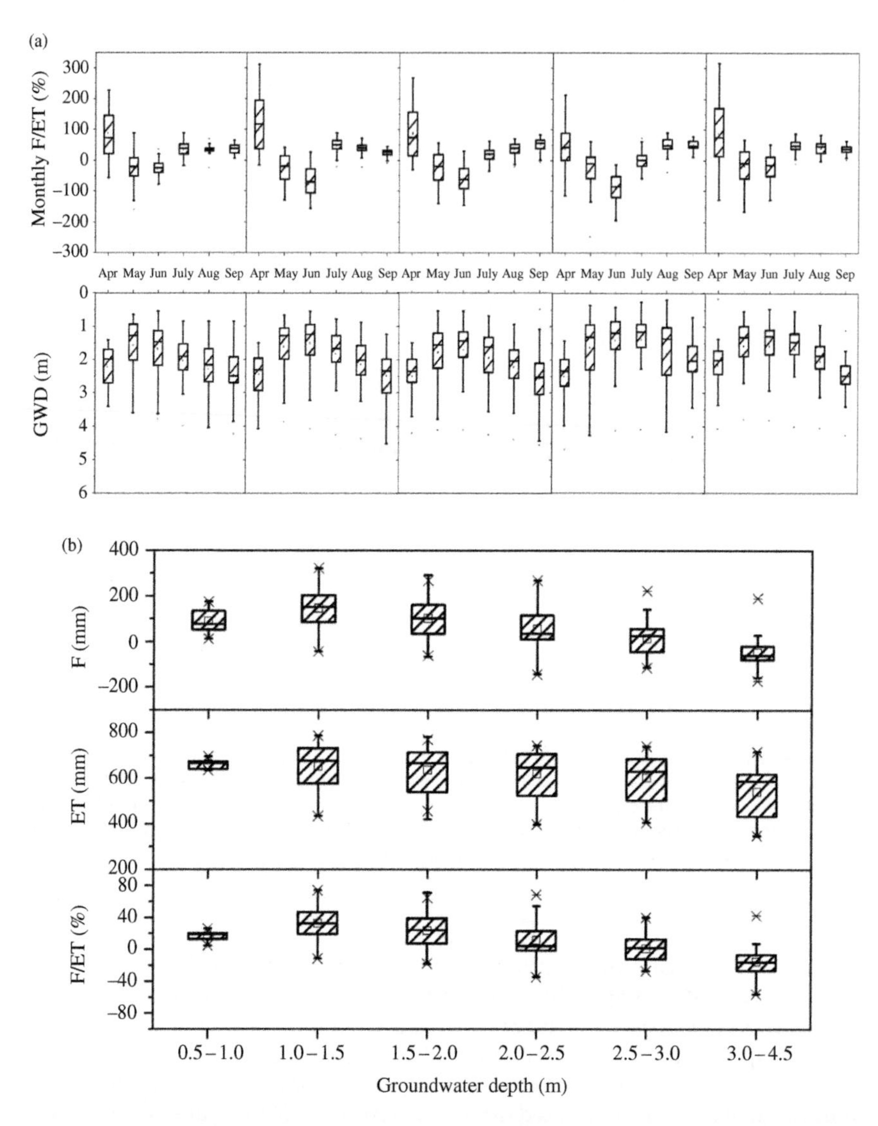

Figure 6.9 (a) Relationship between monthly *F*/ET and groundwater depth, (b) seasonal *F*, ET, *F*/ET, and groundwater depth using 5-yr simulation results (2009–2013). "+" refers to the groundwater capillary rise to soil water more than the soil water percolating to groundwater during crop growing season; "–" refers to groundwater capillary rise to soil water less than the soil water percolating to groundwater during the crop growing season.

upward flux under freezing and thawing and high wind speed in spring leading to high evaporation.

The *F*, ET, and *F*/ET changing with groundwater depth over the crop growing period were statistically analyzed (Figure 6.9b). The groundwater upward flux obviously decreased with groundwater depth from groundwater of 1.0–4.5 m at the regional scale. The soil water content of 90 cm under regions with groundwater depth of 2.5–4.5 m is 50 cm less than that under regions with groundwater depth of 0.5–1.5 m, which confirmed the above result. When the groundwater depth was greater than 2.5–3.0 m, the groundwater upward flux varied from positive values to negative values, which showed that the groundwater contributing to crop growth was less than the soil water percolating to groundwater. Exceptionally, when the groundwater depth was between 0.5 and 1.0 m, the groundwater upward flux was less than that with groundwater depth of 1.0 to–1.5 m. This was due to the low soil saturated conductivity and low groundwater evaporation in this area with a groundwater depth of 0.5–1.0 m. Based on experimental and numerical models, previous research has concluded that shallower groundwater can produce greater groundwater upward flux in arid and semi-arid regions (Grismer & Gate, 1988). Based on lysimetric experiments, Ghamarnia et al. (2013) reported that the groundwater contribution decreased from 65 to 38% when groundwater depth decreased from 0.60 to 1.10 m.

Compared with the trend of *F*, the crop ET decreased slightly with deeper groundwater. When the groundwater depth ranged from 1.0 to 3.0 m, crop ET was approximately 690 mm. However, the average crop ET during the groundwater depths of 0.5–1.0 m were 660 mm, which was attributed to less groundwater upward flux in this region. Similar to the findings of Luo and Sophocleous (2010), ratios of seasonal groundwater evaporation to seasonal potential ET were plotted against the depth to the water table (Figure 6.9). When the groundwater was deeper than 3.0 m, the crop ET decreased significantly due to a lack of groundwater upward flux to crop growth under deeper groundwater. This result could be attributed to lower groundwater upward flux to soil leading to low soil water content in the deep groundwater district mentioned in the previous description.

The effect of groundwater depth on the groundwater contribution to crop ET (*F*/ET) is shown in Figure 6.9b. The variation trend was consistent with the tendency of groundwater upward flux at the regional scale. With groundwater depths of 1.0–4.5 m, the *F*/ET obviously decreased with the decline of groundwater levels. With the groundwater depths of 1.0–1.5 m, the *F*/ET varied up to 65%. These results are in agreement with the findings of previous studies, and the contribution of groundwater to ET increased with a rising water table and decreased from 90 to 7% when the water table depth increased from 50 to 150 cm for the silt clay loam with lysimeters experiments of constant water table depths in the greenhouse (Kahlown et al., 2005; Karimov et al., 2014; Torres & Hanks, 1989). When the groundwater depth is greater than 3.0 m, the groundwater cannot rise through capillary action to the root zone, but some irrigation water can still percolate to the groundwater. Similarly, Luo and Sophocleous (2010) investigated the extinction depth of groundwater evaporation that was approximately 3.8 m. In addition, the effect of the groundwater level on groundwater upward flux is also related to the soil texture, crop growing season, and climate conditions. The irrigation can also significantly affect the groundwater upward flux. Furthermore, the groundwater depth can change the efficiency of irrigation water to some extent, which is due to a part of the deep seepage irrigation water being able to be reused.

Relationship between Water Productivity and Groundwater Depth under Various Irrigation Amounts

Spatial Distribution of WP and IWP

Influenced by crops, soil conditions, groundwater depth, and agricultural practices including fertilization and atmospheric factors, WP varies both spatially and temporarily (Hatfield et al., 2001; Cox et al., 2002). In the 5 yr, the WP slightly fluctuated from 15.0 to 22.2 kg ha^{-1} mm^{-1} at the regional scale. Specifically, the WP in the southeastern part was greater than that in other parts of the study area. This could be attributed to the fact that the average groundwater depth of approximately 2.5 m in the southeastern part (Qinghui area) was the appropriate groundwater depth for WP. With the application of water-saving measures, the WP gradually increased with WP values for 2009–2013 of 18.3, 17.9, 19.0, 19.7, and 20.2 kg ha^{-1} mm^{-1}, respectively. In agreement with the study of Huo et al. (2012), the WP was mainly affected by irrigation amounts and groundwater depth in this relatively dry region.

Irrigation water productivity (IWP) is the ratio of crop yield to the actual irrigation amount. The IWP ranged from 10.4 to 87.3 kg ha^{-1} mm^{-1} in the 5 yr. And due to greater rainfall and lower irrigation in 2012, the average IWP of 32.2 kg ha^{-1} mm^{-1} in 2012 was less than the average IWP of 32.7, 38.4, 33.6, and 43.6 kg ha^{-1} mm^{-1} in other 4 yr. Different from the spatial distribution of WP, the IWP in the northeastern part (Yangjiahe and Huangji areas) was greater than that in the other parts.

Relationship between Water Productivity and Groundwater Depth

Based on the statistical analysis of data for 51 HRUs from 2009 to 2013, the WP under various groundwater depths and irrigation amounts at the regional scale is shown in Table 6.5. Though there was little difference of WP at the regional scale, the mean WP was related to the distribution of groundwater depth with a slightly parabolic trend. Overall, when the groundwater depth was less than 3.0 m, the WP increased with increasing groundwater depth, and then the WP began to decrease with groundwater depths greater than 3.0 m. The average WP at 0.5–1.0, 2.5–3.0, and 3.0–4.5 m during 2009–2013 were 18.7, 19.1, and 19.0 kg ha^{-1} mm^{-1}, respectively, which illustrated that invalid groundwater evaporation that groundwater upward flux unused by crop growing existed under shallower groundwater and water stress affected crop transpiration under deeper groundwater with lower soil water content. On the other hand, with irrigation water increasing, the WP decreased gradually. Thus, simulations seem to agree with the earlier findings of Huo et al. (2012) and Gao et al. (2017), where with appropriate irrigation, the relationship between WP and groundwater depth had a parabolic trend. In agreement with Mueller et al. (2005), the WUE of wheat could be enhanced with deeper groundwater tables to some extent.

Relationship between Irrigation Water Productivity and Groundwater Depth

The IWP under various groundwater depths and irrigation amounts at the regional scale from 2009 to 2013 is shown in Table 6.5. The results showed that the IWP decreased with increasing irrigation amounts for each groundwater depth. The mean IWP decreased from 58.3 to 17.7 kg ha^{-1} mm^{-1}, whereas the irrigation amount increased from 100 to 900 mm. In agreement with the findings of Howell

Table 6.5 Water productivity and irrigation water productivity under various groundwater depths and irrigation amount (kg ha^{-1} mm^{-1})

	Irrigation amount, mm	Groundwater depth, m						
		0.5–1.0	1.0–1.5	1.5–2.0	2.0–2.5	2.5–3.0	3.0–4.5	Mean
Water productivity	100–300	18.7 (±0.4)	19.2 (±1.2)	19.3 (±1.6)	19.2 (±1.9)	20.2 (±1.1)	19.8 (±00.4)	19.4 (±0.5)
	300–500	18.6 (±1.3)	18.9 (±1.4)	19.0 (±1.4)	19.1 (±1.5)	18.7 (±1.6)	19.0 (±1.0)	18.9 (±0.2)
	500–700		17.7 (±1.8)	18.9 (±1.8)	18.9 (±1.4)	18.7 (±1.2)	18.3 (±1.8)	18.7 (±0.2)
	700–900			18.6 (±1.3)	18.7 (±1.4)	18.7 (±1.0)		18.7 (±0.1)
	Mean	18.7 (±0.1)	18.9 (±0.3)	19.0 (±0.3)	19.0 (±0.2)	19.1 (±0.7)	19.0 (±0.7)	
Irrigation water productivity	100–300	58.1 (±13.2)	58.0 (±10.3)	60.8 (±12.6)	62.1 (±16.2)	51.3 (±12.3)	44.2 (±8.1)	58.3 (±6.7)
	300–500	37.6 (±19.0)	37.0 (±9.0)	39.5 (±14.1)	36.1 (±8.2)	35.2 (±10.7)	31.6 (±5.5)	36.2 (±2.6)
	500–700		26.3 (±3.9)	26.4 (±5.2)	29.2 (±9.0)	25.2 (±3.4)	19.7 (±4.1)	25.4 (±3.4)
	700–900			17.7 (±4.0)	18.2 (±1.1)	17.2 (±1.1)		17.7 (±0.5)
	Mean	47.9 (±14.0)	40.4 (±15.8)	36.1 (±18.7)	36.4 (±18.6)	32.2 (±14.6)	31.8 (±12.2)	

et al. (1997) and Huang et al. (2005), the irrigation WUE for biomass and grain yield decreased with increasing irrigation. Alternatively, the effect of groundwater depth on IWP was generally linear. The IWP decreased with the decline of the groundwater table under the same irrigation application, which was due to more groundwater upward flux to crop growing in the shallow groundwater district and less groundwater upward flux with deep groundwater. Similar to the previous study of Huo et al. (2012), the IWP significantly decreased with the decline of the groundwater table under the same irrigation water application, attributable to the shallow groundwater contributing to crop water use.

Conclusions

With a nonweighing lysimeter experiment, the AWPM-SG model was developed to estimate the water fluxes, deep percolation, and capillary rise under different groundwater depth in the experiment and designed scenarios. The model was calibrated and validated using the experimental data in 2007 and 2008, respectively. The simulation of soil moisture, groundwater depth, and LAI resulted in good agreement.

Based on the AWPM-SG model, the spatial distributions of groundwater upward flux, crop ET, crop yield, and WP in the JFZIA from 2009 to 2013 were simulated. Furthermore, the regional groundwater contribution to crop water use and efficiency of irrigation water were quantified from the regional simulation results.

Groundwater contributions to ET can greatly increase biomass productivity. The maximum WP appeared in the district with groundwater depths of approximately 3.0 m. Meanwhile, the distributions of IWP were correlated with the distribution of groundwater depth. Overall, for enhancing the efficiency of irrigation water and IWP, irrigation schedule needs to be optimized by considering groundwater contributions to ET in shallow groundwater fields in arid and semi-arid areas. Furthermore, the contribution varies with other factors, including soil texture and climate condition. In future studies, researchers should further investigate the impact of multiple factors on shallow groundwater contributions to agricultural WP.

Acknowledgments

This research was supported by China's 13th 5-yr key research and development plan (2017YFC0403301), National Natural Science Fund of China (51679236, 51639009, 51809142), and the Second Tibetan Plateau Scientific Expedition and Research Program (STEP) (2019QZKK0207). The contributions of the editor and anonymous reviewers whose comments and suggestions significantly improved this chapter are also appreciated.

Abbreviations

AWPM-SG, Agricultural Water Productivity Model for Shallow Groundwater; DSSAT, Decision Support System for Agrotechnology Transfer; EPIC, Environmental Policy Integrated Climate; ET, evapotranspiration; ET_0, reference evapotranspiration; GSPAC, groundwater–soil–plant–atmosphere continuous; HI, harvest index; HRU, hydrological response unit; IWP, irrigation water productivity; JFZIA, Jiefangzha Irrigation Area; LAI, leaf area index; MRE, mean relative error; NSE, Nash and Sutcliffe model efficiency; PHU, potential heat units; RMSE, root mean square error; SUCROS, Simple and Universal Crop Growth Simulator; SWAP, soil–water–atmosphere–plant; TDR, time domain reflectometry; WIPE, Watershed Irrigation Potential Estimation; WOFOST, World Food Studies; WP, water productivity; WUE, water use efficiency.

References

Abbaspour, K. C., Johnson, A., & van Genuchten, M. T. (2004). Estimating uncertain flow and transport parameters using a sequential uncertainty fitting procedure. *Vadose Zone Journal*, *3*, 1340–1350. https://doi.org/10.2136/vzj2004.1340

Allen, R. Pereira, L., Raes, D., & Smith, M. (2006). *Crop evapotranspiration*. FAO Irrigation and Drainage Paper no. 56.

Bargahei, K., & Mosavi, S. A. A. (2006). Effects of shallow water table and groundwater salinity on contribution of groundwater to evapotranspiration of safflower (*Carthamus tinctorius* L.) in greenhouse. (In Persian.). *Science and Technology of Agriculture and Natural Resources*, *10*, 59–70.

Bouman, B. A. M. (2007). A conceptual framework for the improvement of crop water productivity at different spatial scales. *Agricultural Systems*, *93*, 43–60. https://doi.org/10.1016/j.agsy.2006.04.004

Brown, R. A., & Rosenberg, N. J. (1999). Climate change impacts on the potential productivity of corn and winter wheat in their primary United States growing regions. *Climate Change*, *41*, 73–107.

Cai, L. G., Mao, Z., Fang, S. X., & Liu, H. S. (2003). The Yellow River basin and case study areas. In L. S. Pereira, L. G. Cai, A. Musy, & P. S. Minhas (Eds.), *Water saving in the Yellow River basin: Issue and decision support tools in irrigation* (pp. 13–34). China Agricultural Press.

Campbell, G. S., & Norman, J. M. (1998). *An introduction to environmental biophysics* (2nd ed.). Springer–Verlag.

Carbone, G. J., Mearns, L. O., Mavromatis, T., Sadler, E. J., & Stooksbury, D. (2003). Evaluating CROPGRO-soybean performance for use in climate impact studies. *Agronomy Journal*, *95*, 537–544. https://doi.org/10.2134/agronj2003.5370

Cox, J. W., McVicar, T. R., Reuter, D. J., Wang, H., Cape, J., & Fitzpatrick, R. W. (2002). Assessing rainfed and irrigated farm performance using measures of water use efficiency. In T. R. McVicar, L. Rui, J. Walker, R. W. Fitzpatrick, & L. Chanming (Eds.), *Regional water and soil assessment for managing sustainable agriculture in China and Australia* (pp. 70–81). Australian Centre for International Agricultural Research.

Dalin, C., Qiu, H., & Hanasaki, N. (2015). Balancing water resources conservation and food security in China. *Proceedings of the National Academy of Sciences of the USA*, *112*, 4588–4593.

De Wit, C. T., Brouwer, R., & Penning de Vries, F. W. T. (1970). The simulation of photosynthetic systems. In I. Setlik (Ed.), *Proceedings of the International Biological Program/Plant Production Technical Meeting, Trebon* (pp. 47–70). PUDOC. https://edepot.wur.nl/198106

Easterling, E. W., Crosson, P. R., Rosenberg, N. J., & Lemon, K. M. (1993). Agricultural impacts of and responses to climate change in the Missouri–lowa–Nebraska–Kansas (MINK) region. *Climatic Change*, *24*, 23–61. https://doi.org/10.1007/bf01091476

Gao, X. Y., Bai, Y. N., Huo, Z. L., Huang, G. H., Xia, Y. H., & Steenhuis, S. T. (2017). Deficit irrigation enhances contribution of shallow groundwater to crop water consumption in arid area. *Agricultural Water Management*, *185*, 116–125.

Ghamarnia, H., Golamian, M., Sepehri, S., Arji, I., & Norozpour, S. (2013). The contribution of shallow groundwater by safflower (*Carthamus tinctorius* L.) under high water table conditions, with and without supplementary irrigation. *Irrigation Science*, *31*, 285–299.

Grismer, M. E., & Gate, T. K. (1988). Estimating saline water table contribution to crop water use. *California Agriculture*, *42*, 23–24.

Hatfield, J. L., Sauer, T. J., & Prueger, J. H. (2001). Managing soils to achieve greater water use efficiency: A review. *Agronomy Journal*, *93*, 271–280. https://doi.org/10.2134/agronj2001.932271x

Howell, T. A., Schneider, A. D., & Evett, S. R. (1997). Subsurface and surface microirrigation of corn: Southern High Plains. *Transactions of the American Society of Agricultural Engineers*, *40*, 635–641. https://doi.org/10.13031/2013.21322

Huang, Y. L., Chen, L. D., Fu, B. J., & Gong, J. (2005). The wheat yields and water-use efficiency in the Loess Plateau: Straw mulch and irrigation effects. *Agricultural Water Management*, *72*, 209–222.

Huo, Z. L., Feng, S. Y., Huang, G. H., Zheng, Y. Y., Wang, Y. H., & Guo, P. (2012). Effect of groundwater level depth and irrigation amount on water fluxes at the groundwater table and water use of wheat. *Irrigation and Drainage, 61*, 348–356.

Jaksa, W., & Sridhar, V. (2015). Effect of irrigation in simulating long-term evapotranspiration climatology in a human-dominated river basin system. *Agricultural and Forest Meteorology, 200*, 109–118. https://doi.org/10.1016/j.agrformet.2014.09.008

Jones, J., Hoogenboom, G., Porter, C. H., Boote, K. J., Batchelor, W. D., Hunt, L. A., Wilkens, P. W., Singh, U., Gijsman, A. J., & Ritchie, J. T. (2003). The DSSAT cropping system model. *European Journal of Agronomy, 18*, 235–265. https://doi.org/10.1016/S1161-0301(02)00107-7

Kahlown, M. A., Ashraf, M., & Zia-ul-Haq (2005). Effect of shallow groundwater table on crop water requirement and crop yields. *Agricultural Water Management, 76*, 24–35. https://doi.org/10.1016/j.agwat.2005.01.005

Karimov, A. H., Simunek, J., Hanjra, M. A., Avliyakulov, M. M., & Forkutsa, I. (2014). Effects of the shallow water table on water use of winter wheat and ecosystem health: Implications for unlocking the potential of groundwater in the Fergana Valley (Central Asia). *Agricultural Water Management, 131*, 57–69.

Kendy, E., Gerard-Marchant, P., Walter, M. T., Zhang, Y. Q., Liu, C. M., & Steenhuis, T. S. (2003). A soil-water–balance approach to quantify groundwater recharge from irrigated cropland in the North China Plain. *Hydrological Processes, 17*, 2011–2031. https://doi.org/10.1002/hyp.1240

Kong, F. R. (2009). *The Utilization Efficiency Experiment and Simulation Evaluation of Soil Water and Fertilizer Under Different Groundwater Buried Depth* [Master's thesis, Inner Mongolia Agricultural University]. https://www.dissertationtopic.net/doc/518484

Li, H. S., Wang, W. F., Zhang, G. B., Zhang, Z. M., & Wang, X. W. (2011). GSPAC water movement in extremely dry area. *Journal of Arid Land, 3*, 141–149.

Liu, J. G., Wiberg, D., Zehnder, A. J. B., & Yang, H. (2007). Modelling the role of irrigation in winter wheat yield, crop water productivity, and production in China. *Irrigation Science, 26*, 21–23.

Luo, Y., & Sophocleous, M. (2010). Seasonal groundwater contribution to crop-water use assessed with lysimeter observations and model simulation. *Journal of Hydrology, 389*, 325–335.

Mearns, L. O., Mavromatis, T., Tsvetsinskaya, E., Hays, C., & Easterling, W. (1999). Comparative response of EPIC and CERES crop models to high and low resolution climate change scenarios. *Journal of Geophysical Research, 104*, 6623–6646.

Moriasi, D. N., Arnold, J. G., Van Liew, M. W., Bingner, R. L., Harmel, R. D., & Veith, T. L. (2007). Model evaluation guidelines for systematic quantification of accuracy in watershed simulations. *Transaction of the American Society of Agricultural and Biological Engineers, 50*, 885–900. https://doi.org/10.13031/2013.23153

Morison, J., Baker, N., Mullineaux, P., & Davies, W. J. (2008). Improving water use in crop production. *Philosophical Transactions of the Royal Society B, 363*, 639–658. https://doi.org/10.1098/rstb.2007.2175

Mualem, Y. (1976). A new model for predicting the hydraulic conductivity of unsaturated porous media. *Water Resources Research, 12*, 513–522.

Mueller, L., Behrendt, A., Schalitz, G., & Schindler, U. (2005). Above ground biomass and water use efficiency of crops at shallow groundwater tables in a temperate climate. *Agricultural Water Management, 75*, 117–136.

Niu, X. Z., Esterling, W., Hays, C. J., Jacobs, A., & Mearns, L. (2009). Reliability and input-data induced uncertainty of the EPIC model to estimate climate change impact on sorghum yields in the U.S. Great Plains. *Agriculture, Ecosystems, and Environment, 129*, 268–276. https://doi.org/10.1016/j.agee.2008.09.012

Raes, D., Steduto, P., Hsiao, T. C., & Fereres, E. (2012). *Aquacrop reference manual*. FAO, Land and Water Division.

Ren, D. Y., Xu, X., Hao, Y. Y., & Huang, G. H. (2016). Modeling and assessing field irrigation water use in a canal system of Hetao, Upper Yellow River Basin: Application to maize, sunflower and watermelon. *Journal of Hydrology, 532*, 122–139.

Rosenberg, N. J., Mckenney, M. S., Esterling, W. E., & Lemon, K. M. (1992). Validation of EPIC model simulations of crop responses to current climate and CO_2 conditions: Comparisons with census, expert judgement and experimental plot data. *Agricultural and Forest Meteorology, 59*, 35–51.

Rosenzweig, C., Elliott, J., Deryng, D., Ruane, A. C., Müller, C., Arneth, A., Boote, K. J., Folberth, C., Glotter, M., Khabarov, N., Neumann, K., Piontek, F., Pugh, T. A. M., Schmid, E., Stehfest, E., Yang, H., & Jones, J. W. (2014). Assessing agricultural risks of climate change in the 21st century in a global gridded crop model intercomparison. *Proceedings of the National Academy of Sciences of the USA, 111*, 3268–3273.

Salah, E. E., Maher, A. K., Nasser, A. A., & Urs, S. (2014). Optimal coupling combination between the irrigation rate and glycine betaine levels for improving yield and water use efficiency of drip-irrigated maize grown under arid conditions. *Agricultural Water Management, 140*, 69–78. https://doi.org/10.1016/j.agwat.2014.03.021

Saleh, A., Steenhuis, T., & Walter, M. (1989). Groundwater table simulation under different rice irrigation practices. *Journal of Irrigation and Drainage Engineering, 115*, 530–544.

Sau, F., Boote, K. J., Bostick, W. M., Jones, J. W., & Mínguez, M. I. (2004). Testing and improving evapotranspiration and soil water balance of the DSSAT crop models. *Agronomy Journal, 96*, 1243–1257. https://doi.org/10.2134/agronj2004.1243

Schoups, G., Hopmans, J., Young, C., Vrugt, J. A., Wallender, W. W., Tanji, K. K., & Panday, S. (2005). Sustainability of irrigated agriculture in the San Joaquin Vally, California. *Proceedings of the National Academy of Sciences of the USA, 102*, 15352–15356.

Simunek, J., van Genuchten, M., & Sejia, M. (2005). *The HYDRUS-1D Software Package for Simulating the One-dimensional Movement of Water, Heat and Multiple Solutes in Variability-Saturated Media.* Development of Environment Sciences/University of California.

Steenhuis, T. S., & Van Der Molen, W. H. (1986). The Thornthwaite–Mather procedure as a simple engineering method to predict recharge. *Journal of Hydrology, 84*, 221–229.

Sun, H. Y. (2014). *The analysis of different scales diversity law of irrigation water efficiency and water saving potential in hetao irrigation area of inner Mongolia.* Inner Mongolia Agricultural University.

Talebizadeh, M., & Moridnejad, A. (2011). Uncertainty analysis for the forecast of lake level fluctuation using ensembles of ANN and ANFIS models. *Expert Systems with Applications, 38*, 4126–4135.

Torres, J. S., & Hanks, R. J. (1989). Modeling water table contribution to crop evapotranspiration. *Irrigation Science, 10*, 265–279.

van Dam, J. C., Groenendijk, P., Hendriks, R. F. A., & Kroes, J. G. (2008). Advances of modeling water flow in variably saturated soils with SWAP. *Vadose Zone Journal, 7*, 640–653. https://doi.org/10.2136/vzj2007.0060

van Diepen, C. A., Wolf, J., van Keulen, H., & Rappoldt, C. (1989). WOFOST: A simulation model of crop production. *Soil Use and Management, 5*, 16–24.

van Genuchten, M. T. (1980). A closed-form equation for predicting the hydraulic conductivity of unsaturated soils. *Soil Science Society of America Journal, 44*, 892–898. https://doi.org/10.2136/sssaj1980.03615995004400050002x

Wang, X. C., & Li, J. (2010). Evaluation of crop yield and soil water estimates using the EPIC model for the Loess Plateau of China. *Mathematical and Computer Modelling, 51*, 1390–1397.

Wang, X. S., Yue, W. F., & Yang, J. Z. (2004). Analysis on water cycling in GSPAC system of Tetao-irigation district, in Inner Mongolia, China. *Journal of Irrigation and Drainage, 23*, 30–33. https://doi.org/10.1007/BF02873091

Wang, X. W., Huo, Z. L., Feng, S. Y., Guo, P., & Guan, H. D. (2016). Estimating groundwater evapotranspiration from irrigated cropland incorporating root zone soil texture and moisture dynamics. *Journal of Hydrology, 543*, 501–509. https://doi.org/10.1016/j.jhydrol.2016.10.027

Williams, J. R. (1995). The EPIC model. In V. P. Singh (Ed.), *Computer Models of Watershed Hydrology* (pp. 909–1000). Water Resources Publications.

Williams, J. R., Wang, E., Meinardus, A., Harman, W. L., Siemers, M., & Atwood, J. D. (2006). *EPIC users guide v. 0509*. Texas A&M University.

Williams, J. R., Jones, C. A., Kiniry, J. R., & Spanel, D. A. (1989). The EPIC crop growth model. *Transactions of the American Society of Agricultural Engineers, 32*, 497–511.

Wit, K. E. (1967). *Apparatus for measuring hydraulic conductivity of undisturbed soil samples*. Bulletin Institute for Land and Water Management Research.

Xu, X., Huang, G. H., Qu, Z. Y., & Pereira, L. S. (2010). Assessing the groundwater dynamics and impacts of water saving in the Hetao Irrigation District, Yellow River Basin. *Agricultural Water Management, 98*, 301–313.

Xu, X., Sun, C., Qu, Z. Y., Huang, W. Z., Ramos, T. B., & Huang, G. H. (2015). Groundwater recharge and capillary rise in irrigated areas of the upper Yellow River Basin assessed by an agro-hydrological model. *Irrigation and Drainage, 64*, 587–599. https://doi.org/10.1002/ird.1928

Yang, J., Wan, S., Deng, W., & Zhang, G. (2007). Water fluxes at a fluctuating groundwater table and groundwater contributions to wheat water use in the lower Yellow River flood plain, China. *Hydrological Processes, 21*, 717–724.

Zhou, A. G., Ma, R., & Zhang, C. (2005). Vertical water cycle and its ecological effect in inland basins, Northwest China. (In Chinese, with English abstract.). *Advances in Water Science, 16*, 127–133.

7

Seungtaek Jeong, Han-Yong Kim,
Jonghan Ko, and Byunwoo Lee

Use of Data and Models in Simulating Regional and Geospatial Variations in Climate Change Impacts on Rice and Barley in the Republic of Korea

Abstract

Global warming can impact crop productivity across climates, land-orms, and soils differently due to geospatial variations in factors that influence crop growth. In this chapter, we report our efforts to use data and crop models for simulation of regional and geographical variations in barley (*Hordeum vulgare* L.) and rice (*Oryza sativa* L.) yields in the Republic of Korea (ROK) in future environments under climate change. Barley (2014–2016) and rice (2009–2010) were grown in a temperature gradient field chamber (TGFC) system at Chonnam National University, Gwangju, Republic of Korea (ROK). Greenhouse gasses emission trajectories applied to simulate the impacts of climate change on the crops were an A1B scenario for rice and representative concentration pathway (RCP) 4.5 and RCP 8.5 scenarios for barley. These crops were simulated using the CERES-Barley 4.6 and CERES-Rice 4.0 models. A geospatial crop simulation modeling (GCSM) system was also developed using the CERES-Barley model. Experimental and simulation results showed that CO_2 fertilization effects on dry matter assimilation of rice were negated by the yield reductions due to temperature increases, whereas CO_2 effects on barley dominated the temperature effects, except for in a few areas. Rice production increased at higher latitudinal regions due to positive temperature effects. Using a *k* means clustering the impact of climate change on barley yield, the whole

Enhancing Agricultural Research and Precision Management for Subsistence Farming by Integrating System Models with Experiments, First Edition. Edited by Dennis J. Timlin and Saseendran S. Anapalli.

DOI: 10.1002/9780891183891.ch07

country was classified into six categories. The CERES models demonstrated the capability to be utilized at a local-to-regional scale to investigate the effects of climate change on barley and rice production under a monsoonal climate system. The GCSM system could be utilized to simulate geographical variations of the effects of climate change on crops. Further development of the system could allow to pursue possible interpretations of future food supply uncertainty.

Introduction

Crop growth is mainly attributable to genetic factors and integrated responses of various eco-physiological processes to environmental conditions. Solar radiation, temperature, CO_2 concentration, nutrients, water, and field management influence crop performance. Although there have been great scientific efforts in field experiments to examine the potential effects of these factors on agricultural production, they cannot elucidate all the environmental variables and their interactions. In this context, cropping system models that are adequately calibrated and validated can be used to examine the combined effects of various chemical, physical, and biological processes (Ahuja et al., 2000; Kirschbaum, 2000). Crop models frequently used to simulate the growth and development of crops include Crop Environment Resource Synthesis (CERES) models (IBSNAT, 1989), CROPGRO (Boote et al., 1998), and SUBSTOR (Griffin et al., 1992) in the Decision Support System for Agrotechnology Transfer (DSSAT) package (Jones et al., 2003), the Environmental Policy Integrated Climate (EPIC) model (Williams et al., 1989), SHOOTGROW (McMaster, 1993), and the World Food Studies (WOFOST) model (Hijmans et al., 1994). Out of these crop models, those in the DSSAT package have been widely used by agronomists and ecologists to explore crop productivities under various environmental conditions and potential scenarios, including climate change (CC).

Global CO_2 concentrations, accounting for 77% of greenhouse gases (GHGs), have increased by approximately 80% from 1970 to 2004. Concentrations will likely rise to a range of 421–936 mg L^{-1} and will be associated with an increase in mean global surface temperature between 2000 and 2100 (IPCC, 2013). Increases in anthropogenic GHGs in the atmosphere have led to an increase in global mean surface temperatures of 0.74 °C ± 0.18 °C over the last 100 yr (1906–2005). The Intergovernmental Panel on Climate Change (IPCC) has opined that the global warming trends from 1986–2005 to 2081–2100 will vary depending on different scenarios of GHG concentration trajectories; for example, the representative concentration pathway (RCP) 2.6, RCP 4.5, RCP 6.0, and RCP 8.5. These project a temperature increase of 0.3–1.7 °C based on RCP 2.6, 1.1–2.6 °C based on RCP 4.5, 1.4–3.1 °C based on RCP 6.0, and 2.6–4.8 °C based on RCP 8.5. This estimated elevation of atmospheric CO_2 and the associated increase in temperature can further influence agricultural production due to changes in evapotranspiration, plant growth rates, plant litter composition, and the nitrogen–carbon cycle (Long et al., 2006). The likely impacts in a given location will vary depending on the magnitude of these changes, responses of particular crops, location-specific management, and socio-economic conditions. Therefore, it is critical to investigate the regional impacts of projected increases in GHGs and subsequent global climate change on the production of crops to recommend appropriate adaptation measures in time.

There have been many scientific efforts to quantify global crop production, considering climate change (Lobell & Field, 2007; Rosenzweig et al., 2014). The global distribution of croplands will be continuously transformed due to contemporary drivers of socio-economic development and environmental fluctuations, and possible influences of climate change (Ramankutty et al., 2002). It is also assumed that local and global crop productivity will vary considerably due to various landscapes (Lobell et al., 2011; Olesen et al., 2011; Thornton et al., 2009). There have been some efforts to reproduce geospatial variations to define effects of climate change on staple crops, such as paddy rice (*Oryza sativa* L.) in Southeast Asia (Chun et al., 2016; Li et al., 2017) and maize (*Zea mays* L.) and bean (*Phaseolus vulgaris* L.) in East Africa (Thornton et al., 2009). However, a crop-modeling system that can project spatiotemporal crop productivity with a fine grid scale is necessary to determine local variations in crop production. Climate change scenarios project that South Korea is vulnerable to climate change, equally likewise or more than any other region and continent (IPCC, 2013). Evaluating the impacts of climate change on crops and finding efficient adaption measures have also been of great interest to agricultural policymakers, scientists, and stakeholders in this country. Various efforts have been made based on experimental field studies (e.g., Kim et al., 2011; Kim & Lee, 2019) and others using crop simulation approaches (e.g., Kim et al., 2013; Ko et al., 2019). Although crop modeling studies have focused on simulating the effects of climate on crop productivity, there have been additional efforts to investigate geographical variation in the productivity influenced by the climate change impacts based on the development of crop modeling systems (Kim et al., 2015, 2017, 2020; Ko et al., 2019). This chapter aims to inform recent ongoing efforts to project geospatial variations in the responses of staple crops to global climate change as a case study in South Korea (Kim et al., 2013, 2017; Ko et al., 2014, 2019).

Simulation of Grain Yields of Barley and Rice under Climate Change

Field Experimental Data

Temperature gradient field chamber (TGFC) systems equipped with CO_2 enrichment can simulate the projected increases in atmospheric CO_2 concentration and air temperature (Kim et al., 2011). The TGFC covered by a highly transparent film had a tunnel shape of 24.0 m in length, 2.4 m in width, and 2.0 m in height to control temperature with a gradient from the air inlet to outlet, being able to control CO_2 based on fumigation. There are a few TGFC systems in South Korea. One of the TGFC systems is located at Chonnam National University (35°10′ N, 126°53′ E; 33 m above sea level), Gwangju (Figure 7.1). The TGFC has been applied to reproduce potential growths and yields of paddy rice and winter barley (*Hordeum vulgare* L.) under experimental climate change regimes (Kim et al., 2013, 2017; Ko et al., 2019).

Three local varieties of paddy rice (Nampyeong, Saegyewha, and Unkwang) were cultivated under the experimental climate change treatments with two CO_2 levels and three temperature levels in the TGFC facility for 2 yr from 2009 to 2010. The two CO_2 levels included ambient (A) CO_2 and elevated (E) CO_2 (650 mg L^{-1}), and the three temperature levels comprised local ambient temperature and raised temperatures (1 and 2 °C above the ambient temperature). The

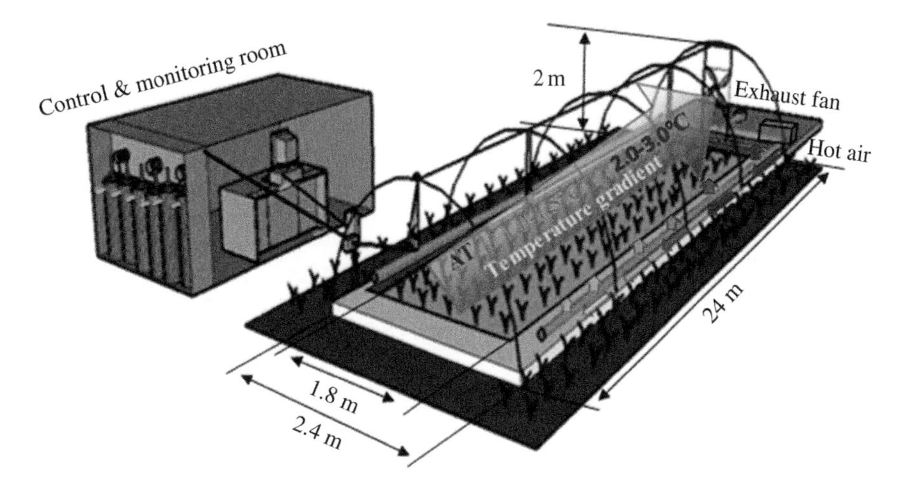

Figure 7.1 Schematic illustration of the temperature gradient field chamber at Chonnam National University, Gwangju, South Korea.

treatment was designed to reproduce future elevated temperature conditions based on climate change regimes of the Special Report on Emission Scenarios (IPCC, 2007).

Four local varieties of barley (DaJin, HeenChal, KeunAl, and SaeChal) were cultivated in an open field, and two cultivars (HeenChal and SaeChal) were grown in the TGFC facility and treated with three temperature levels under the current level of atmospheric CO_2 concentration for three growing seasons from 2014 to 2016 (Figure 7.2). The three temperature levels were local ambient temperature and raised temperatures (1.5 and 3.0 °C above the ambient temperature). The treatment was designed to reproduce future elevated temperature conditions only under the current ambient CO_2 condition based on RCP 4.5 and RCP 8.5 climate change scenarios (IPCC, 2013).

Simulation of Rice and Barley

In this study, both CERES-Rice and CERES-Barley in the DSSAT package were employed in simulation of climate change impacts on rice and barley in South Korea. The CERES-Rice model was used to model measured grain yields grown under the experimental climate change regimes (Figure 7.3). Moreover, the DSSAT–CERES plant growth models (Jones et al., 2003) have been used to simulate yield components, leaf numbers, and phenological stages of the crops. In this section, we report and discuss the recent study results in measured barley and rice yields grown in the TGFC experiments compared with simulated barley and rice yields using CERES-Barley and CERES-Rice models. In both 2009 and 2010, grain yields were higher in ECO_2 than in ACO_2. Grain yields declined with elevated temperatures. The model adequately reproduced grain yields with variations due to the environmental changes and those with the cultivar treatment, as the results were in agreement with the corresponding measured grain yields. Meanwhile, the CERES-Barley model was used to simulate measured grain yields grown under the experimental climate change regimes. Grain

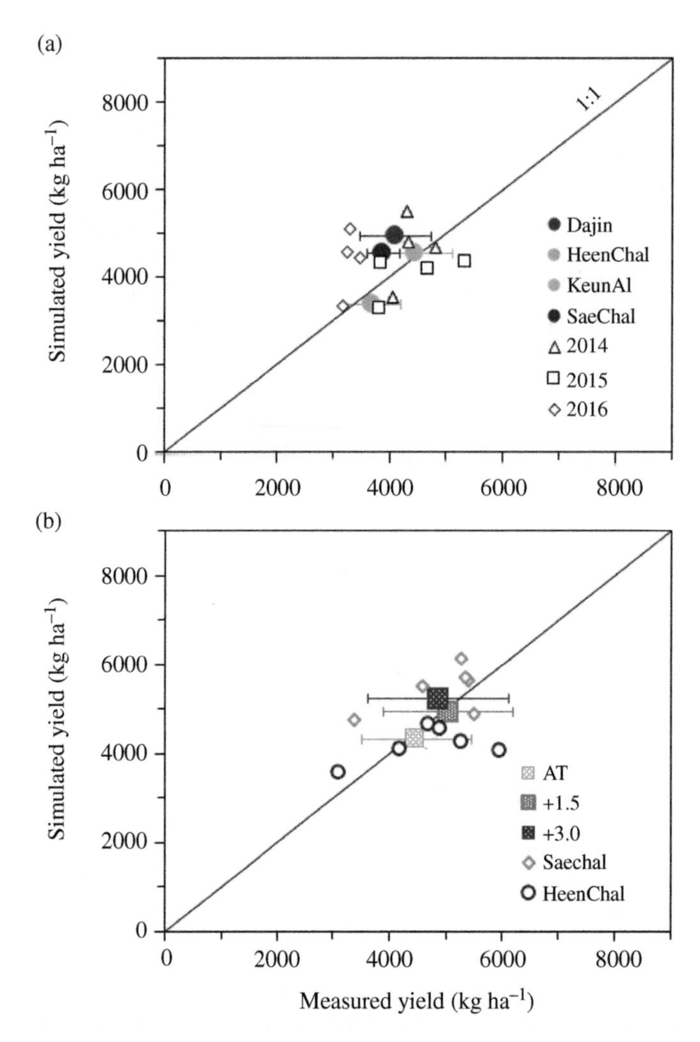

Figure 7.2 Simulated vs. measured grain yields of barley for four local varieties of SaeChal, HeenChal, KeunAl, and Dajin during three crop seasons from (a) 2014 to 2016 and (b) for three temperature regimes of atmospheric temperature (AT), +1.5 °C, and +3.0 °C using two cultivars of SaeChal and HeenChal (Ko et al., 2019). Horizontal bars represent ±1 standard deviation (*n* = 9 for the four variety treatments in Figure 7.3a and *n* = 12 for the three temperature regimes in Figure 7.3b).

yields increased with elevated temperatures. The model simulated grain yields with variations due to the elevated temperatures and those with the cultivar treatment adequately in agreement with the corresponding measured grain yields.

The CERES-Barley and CERES-Rice models were successfully calibrated and validated for simulations of barley and rice productivities grown in the TGFC system at Gwangju, South Korea. It was previously reported that models could

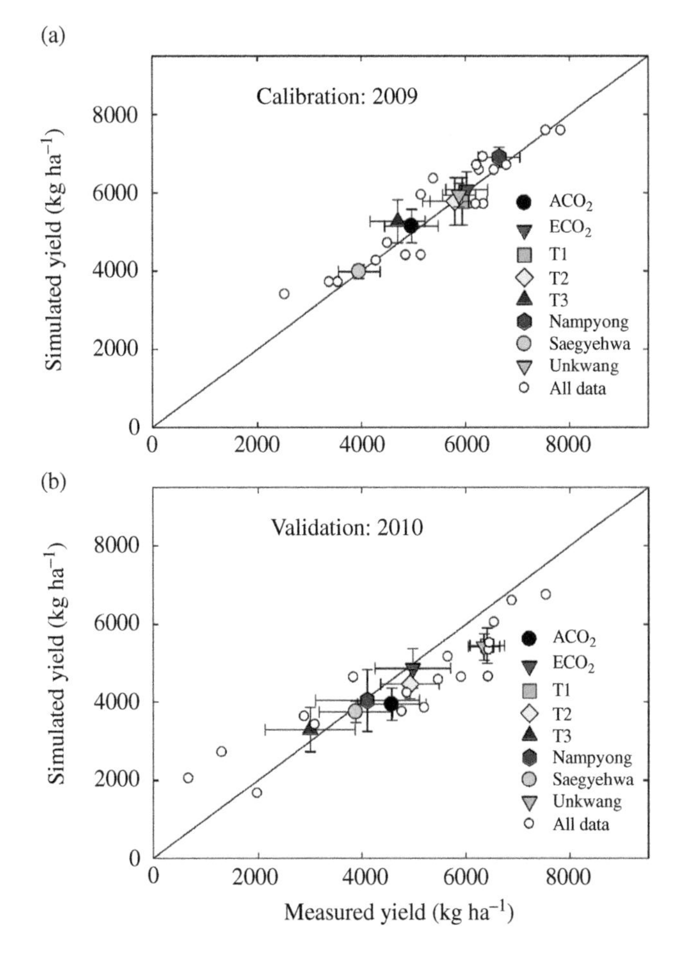

Figure 7.3 Simulated vs. measured yields of three rice varieties of Nampyong, Saegyewha, and Unkwang in (a) calibration and (b) validation using the data obtained from the temperature gradient field chamber experiment (Kim et al., 2013). ACO_2 and ECO_2 represent ambient CO_2 and elevated CO_2 concentrations. T1, T2, and T3 represent ambient temperature and elevated temperatures (1 and 2 °C above the ambient temperature).

simulate CO_2 effects of the free air CO_2 enrichment (FACE) on crop productivity, e.g., those of sorghum (Fu et al., 2016b), soybean (Fu et al., 2016a), and wheat (Asseng et al., 2004; Ko et al., 2010; Tubiello et al., 1999). The recent study results also demonstrated this capability in the context of simulation of barley and paddy rice production under the TGFC system using the CERES-Barley and CERES-Rice models. Although the models proved successful for these study outcomes, we believe that further improvement could be made. These models should be further developed to address more sophisticated issues and field crop research of scientific interest, such as detailed crop development and growth performance, and plant–water–soil–nutrient interactions.

Simulation of the Climate Change Impacts on Barley and Rice

We also employed the CERES-Rice 4.0 model to simulate the potential influences of climate change on rice yield for a 25-yr period, with the target years 2050 (2037–2062) and 2100 (2087–2112) for an A1B GHGs emission scenario for Gwangju, Korea, and the other specified regions. For this purpose, several reliable general circulation models (GCMs) were used to simulate temperature and precipitation projections (Kim et al., 2013). The projected data were based on the A1B scenario with increased radiative forcing for the year 2050 by 550 mg L^{-1} and the year 2100 by 750 mg L^{-1} of atmospheric CO_2 concentration. Each 25-yr GCM projections targeted around 2050 and 2100 were analyzed to account for inter-year climate variability, with their deviations from the recent 25-yr baseline (1986–2010). Then, the GCM projection for temperature, precipitation, and solar radiation were superimposed over the 25-yr baseline using the maximum available climate history obtained from a weather station in Gwangju by the Korea Meteorological Administration. The calculated average temperature increase was then added to the daily minimum and maximum temperatures. Changes in precipitation and solar radiation were also used to adjust the daily rainfall and solar radiation values. These projected climate conditions were used to simulate the effects of CO_2, temperature, and precipitation changes on paddy rice yield using the validated model for the TGFC data set conditions. The CERES-Rice model was also employed to simulate the probable influences of climate change on yield for other short grain (Japonica) rice-producing regions in East Asia for a 12-yr period, at target years of 2050 (2044–2055) and 2100 (2094–2105).

In the case of simulating barley using the CERES-Barley 4.6 model, we adopted the projection of CO_2 concentration for two GHGs concentration trajectory scenarios: RCP 4.5 and RCP 8.5 (IPCC, 2013). The prediction of climate variables included temperature and precipitation. The primary trend was characterized by projected values provided by the Korea Meteorological Administration centered on years 2030, 2050, 2070, and 2100 under RCP 4.5 and RCP 8.5 (Ko et al., 2019). These projected values were simulated using HadGEM2-AO and HadGEM3-RA models according to climate change scenarios obtained from coordinated regional climate downscaling experiment (CORDEX) initiative created by the Task Force for Regional Climate Downscaling (TFRCD) and established by the World Climate Research Programme (WCRP) in 2009. Large-scale climate variables from HadGEM2-AO were dynamically downscaled to a physically consistent evolution on a smaller (0.44° by 0.44°) scale using the HadGEM3-RA model. Further information on these models can be found at the CORDEX–East Asia website (http://cordex-ea.climate.go.kr/cordex/). Baseline regional climate data contained 12-yr (1999–2011) 3-km pixel-by-pixel climate data of South Korea. Climate data were projected using a local climate model (Weather Research and Forecasting model, WRF) to obtain high-resolution regional agro-climate indices based on a dynamical downscaling method (Ahn et al., 2010). Regional shifts estimated from baseline local climate data were added into the primary trend of climate change to produce regional projections of daily climate data. These projected data were used for each 12-yr projection based on the 12-yr baseline of 1999–2011 to include inter-annual climate variability. Projected variations in temperature and precipitation were superimposed on the 12-yr baseline.

Varietal, Local, and Geographical Variations in Grain Yields of Barley and Rice in a Changing Climate

For future environments, the cumulative distribution function (CDF) of simulated yields of three cultivars for the 25 baseline years (BL: 1986–2010) was compared with the projections (i.e., individual effects of CO_2, temperature, precipitation, solar radiation, and the four factors combined) for years centered on 2050 (2037–2062) and 2100 (2087–2112) (Figure 7.4 and Table 7.1). With increasing CO_2 concentrations (i.e., 550 mg L^{-1} for 2050 and 750 mg L^{-1} for 2100), yields (Figure 7.4a–c) were amplified over BL in a significant manner across all rice varieties. However, yields (Figure 7.4d–f) decreased substantially over BL with increasing temperature

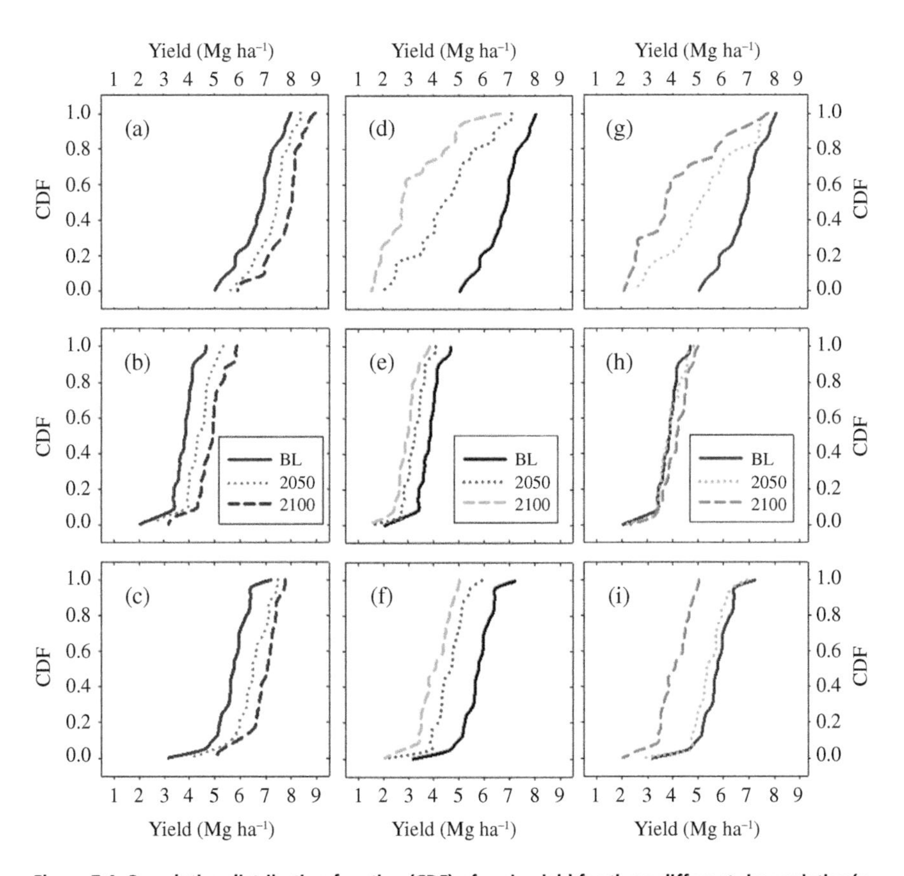

Figure 7.4 Cumulative distribution function (CDF) of grain yield for three different rice varieties (a, d, and g = Nampyeong; b, e, and h = Saegyewha; c, f, and i = Unkwang) cultivated under Gwangju's temperate conditions, comparing the simulated yield for the 25 baseline years (1986–2010) with the projections of CO_2 only (a, b, and c), temperature only (d, e, and f), and the combined effects of CO_2, temperature, precipitation, and solar radiation (g, h, and i) for the years centered on 2050 (2037–2062) and 2100 (2087–2112), modified from Kim et al. (2013).

Table 7.1 Statistical analysis of the future change impacts of CO_2, temperature, and four factors (CO_2, temperature, precipitation, and solar radiation) combined on paddy rice yield for the three different cultivars from the simulation data (Fig. 4).

		CO_2		Temperature		Combined	
Cultivar	Year†	K-S‡ test	CDF mean┘ (kg ha^{-1})	K-S test	CDF mean (mm)	K-S test	CDF mean (kg ha^{-1})
Nampyeong	Baseline	–	6723^{c}	–	6723^{a}	–	6723^{a}
	2050	0.016	7314^{b}	<.001	4518^{b}	0.002	5156^{b}
	2100	0.001	7803^{a}	<.001	3172^{c}	<.001	4120^{c}
Saekyewha	Baseline	–	3763^{c}	–	3763^{a}	–	3763^{a}
	2050	0.001	4348^{b}	0.002	3237^{b}	0.699	3837^{a}
	2100	<.001	4838^{a}	<.001	2910^{c}	0.037	4081^{a}
Unkwang	Baseline	–	5677^{c}	–	5677^{a}	–	5677^{a}
	2050	0.002	6441^{b}	<.001	4576^{b}	0.281	5392^{a}
	2100	<.001	6886^{a}	<.001	3996^{c}	<.001	3996^{b}

† 2050 and 2100 represent the 25-year centers for 2037-2062 and 2087-2112, respectively.
‡ K-S refers to the Kolmogorov-Smirnov test. The numbers in the column represent *p*-values at 95% confidence intervals.
┘ CDF represents cumulative distribution function. The values followed by the same lower-case letters within each column for each treatment are not significantly different based on the Duncan's Multiple Range Test (DMRT) at 95% confidence intervals.

(i.e., 2.08 °C for 2050 and 3.19 °C for 2100) for all cultivars. With regard to the precipitation and solar radiation changes, no significant yield increases or decreases were seen (data not shown). With all four factors combined for 2050 and 2100, yields (Figure 7.4g–i) decreased significantly according to the Duncan's multiple range test, significant at the 95% confidence interval, for both projections of Nampyeong and the 2100 forecast of Unkwang. Here, the CERES-Rice model projected that the effects of CO_2 fertilization on grain yield for future years would produce higher yields in 2050 and 2100 than the baselines for all the cultivars. However, for Nampyeong and Unkwang, it appeared that temperature increases, which significantly diminished yields, dominated the yield increases due to CO_2 elevation. Adams et al. (1990) reported that future changes in temperature and precipitation could lead to improvements in crop water demands and reductions in yield, whereas increased CO_2 enhanced crop yields in US agriculture. The current study suggests that depending on the cultivar selection, the effect of increased temperatures could be positive or negative on paddy rice production in this particular region. Therefore, increased temperatures could moderate the effects of other climate parameters such as CO_2 and precipitation. It was also presumed that CO_2 fertilization effects could dominate over the effects of other climate factors in a cultivar, such as Saekyewha, which showed less sensitivity to temperature changes. This finding is consistent with the report by Saseendran et al. (2000), which concluded that rice yield increases under rain-fed conditions were possible under the projected climate change scenario. In the case of the other cultivars (e.g., Nampyeong and Unkwang), the effects of temperature changes would dominate over the effects of changes in CO_2. Overall, these results implied that various cultivation options, such as cultivar selection, might overcome the potential impacts of climate change on rice production in future years.

The CERES-Rice model was also applied to simulate the potential impacts of climate change on paddy production for other short grain (Japonica) rice-producing regions in East Asia for a 12-yr period, at years centered on 2050 (2044–2055) and 2100 (2094–2105), for an SRES-based A1B GHGs emission scheme (Kim et al., 2013). The effects of climate change on rice yield at different altitudes were also simulated at mountainous highland topographies (Ko et al., 2014). The yields under the combined climate change effect (i.e., CO_2, temperature, precipitation, and solar radiation) at various latitudes decrease in Hiroshima, Japan, similarly to those in Gwangju, Korea, and to increase in the other regions (Figure 7.5). The model simulated the highest yield increases of +24% in 2050 and +29% in 2100 in Harbin, China. CERES-Rice also projected yield increases at mountainous terrains (>235 m) in South Korea in future years (Ko et al., 2014). The yield increases were more significant with increasing latitude and altitude. Meanwhile, the model simulated small increases (<±5%) in yield from 2050 to 2100. The combined effects of CO_2, temperature, precipitation, and solar radiation on yields at diverse latitudinal regions were mainly attributed to those of CO_2 and temperature (Kim et al., 2013). Carbon dioxide fertilization effects were relatively higher in Gwangju, Korea, and in Hiroshima, Japan. The results indicate that temperature effects in these regions would dominate the impact in comparison with the higher latitudinal regions. For the areas in the latitudinal range of approximately 34°–42°, CO_2 fertilization effects would negate the adverse effects of temperature. For those at higher latitudes (i.e.,

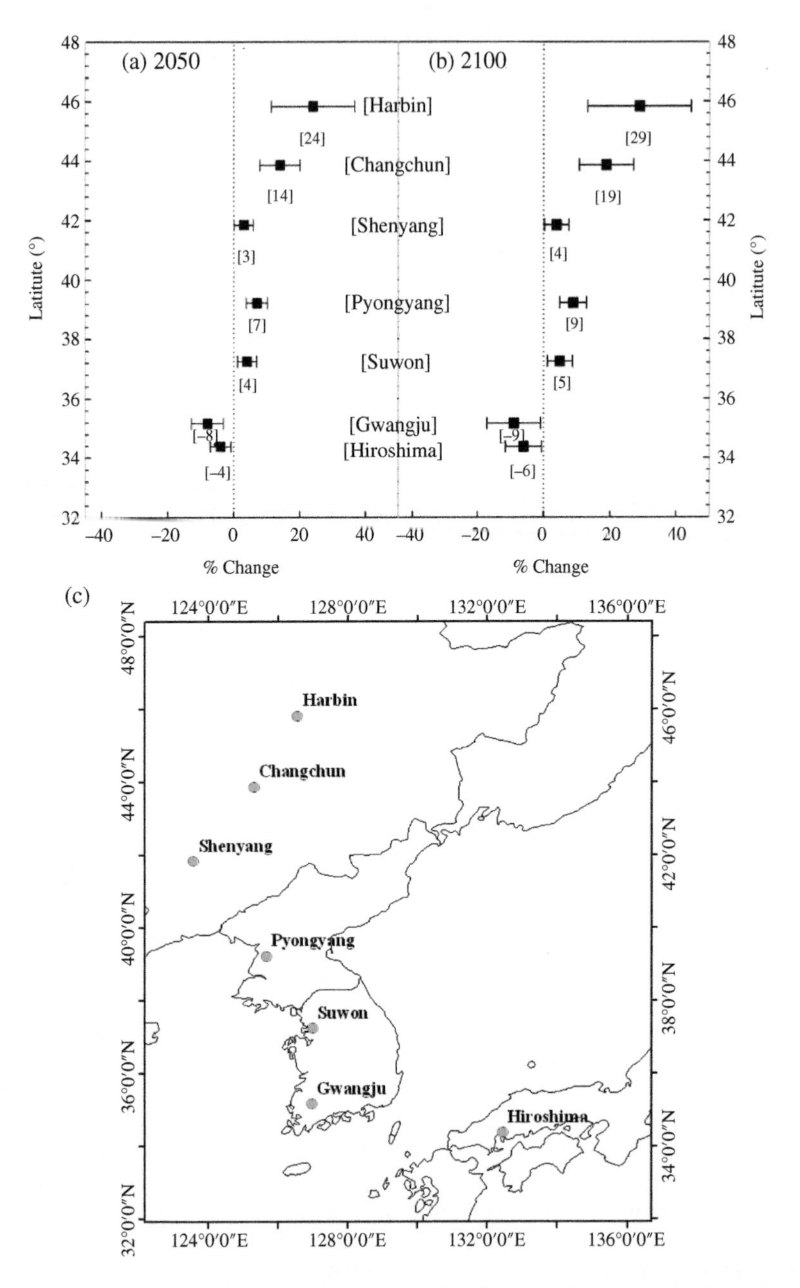

Figure 7.5 Percentage changes in rice yield from the 12 baseline years (1997–2008) for the projected years centered on (a) 2050 (2044–2055) and (b) 2100 (2094–2105), simulated with the combined effects of CO_2, temperature, precipitation, and solar radiation for the various East Asian rice production locations at (c) Harbin, Changchun, and Shenyang, China; Pyongyang, Suwon, and Gwangju, Korea; Hiroshima, Japan. Horizontal bars represent ±1 SE (n = 12). The mean percentage change values of rice yield for each region are given in parentheses, which were determined from the simulated mean yield of the combined effects averaged for three varieties (refer to Table 7.1) in the baseline years, modified from Kim et al. (2013).

>42°), positive effects on yield would be mostly attributable to temperature with small effects from CO_2 fertilization. This suggests that temperatures during the grain formation were somewhat higher than the optimum mean temperature of 25 °C (according to Baker et al., 1995) at the lower latitudinal regions (i.e., <42°), and vice versa at the higher latitudinal regions. These results are consistent with those reported earlier in Japan (Horie et al., 1995) and those recently conveyed in California, USA (Lee et al., 2011). Horie et al. (1995) projected yield increases at higher latitudinal regions, such as Sapporo (+16.6%), and yield decreases at lower latitudinal regions, such as Hiroshima (−15.0%), using a modified SIMRIW crop model simulated under a CO_2 concentration of 700 mg L^{-1} with the GISS model climate. The report by Lee et al. (2011) also showed yield reductions at comparatively lower latitudes (i.e., approximately 36°–40°) using a DAYCENT model under the A2 and B1 scenarios with the six GCMs. Although the current results demonstrated a general resemblance to those presented by Adams et al. (1990) and Hatfield et al. (2011), different climate change projections with maximum and minimum temperature variability might show different results.

A geospatial crop simulation modeling (GCSM) system developed earlier (Kim et al., 2017) was used to simulate the potential impacts of climate change on regional projections of barley yield (Ko et al., 2019). The GCSM simulated that barley grain yields of all four barley cultivars of DaJin, HeenChal, KeunAl, and SaeChal in all of South Korea would increase in future years (2030, 2050, 2070, and 2100) under both RCP 4.5 and RCP 8.5 GCM forcing scenarios (Figure 7.6). Grain yields of all barley cultivars adequately represented spatial variations within geographical regions of the whole country, showing higher yields in southern coastal areas and lower yields in mountainous inner areas. Mean grain yields of all barley cultivars in the future under the RCP 8.5 scenario would vary regionally in 2030 and 2050. They would increase rapidly in 2070 and 2100 for most of the administrative areas and the whole country, except for Jeju. In Jeju, mean grain yields are likely to decrease rapidly in 2070 and 2100. Mean grain yields under the RCP 8.5 scenario were projected to decline by 2.7% by 2030 compared with those at the baseline, and to increase rapidly from 22.3% in 2050 to 42.9% in 2100. For nine administrative regions (Chonbuk, Chonnam, Chungbuk, Chungnam, Jeju, Kangwon, Kyunggi, Kyungbuk, and Kyungnam), mean grain yields were projected to increase with trends similar to those of the whole nation, except for Jeju, although some variations existed due to different meteorological and environmental conditions. In the case of Jeju, mean grain yields were projected to decrease gradually in the future (from +3.4% in 2050 to −26.6% in 2100 under RCP 8.5).

There have been significant efforts to compute global crop production, especially considering the changing climate (Lobell & Field, 2007; Rosenzweig et al., 2014); however, it is not possible at present to measure the collective influence of climate change on global agricultural productivity (Gornall et al., 2010). Difficulties mainly arise from complications in suitable classification of cultivated lands, uncertainties in climate projection models, and genetic, environmental, and regional variabilities in crop production. The global distribution of arable areas is likely to change continuously, not only due to current drivers of socio-economic development and environmental changes, but also due to potential impacts of climate change (Ramankutty et al., 2002). It is also likely that the variability in regional and global crop productivity will significantly change (Lobell et al., 2011; Olesen et al., 2011; Thornton

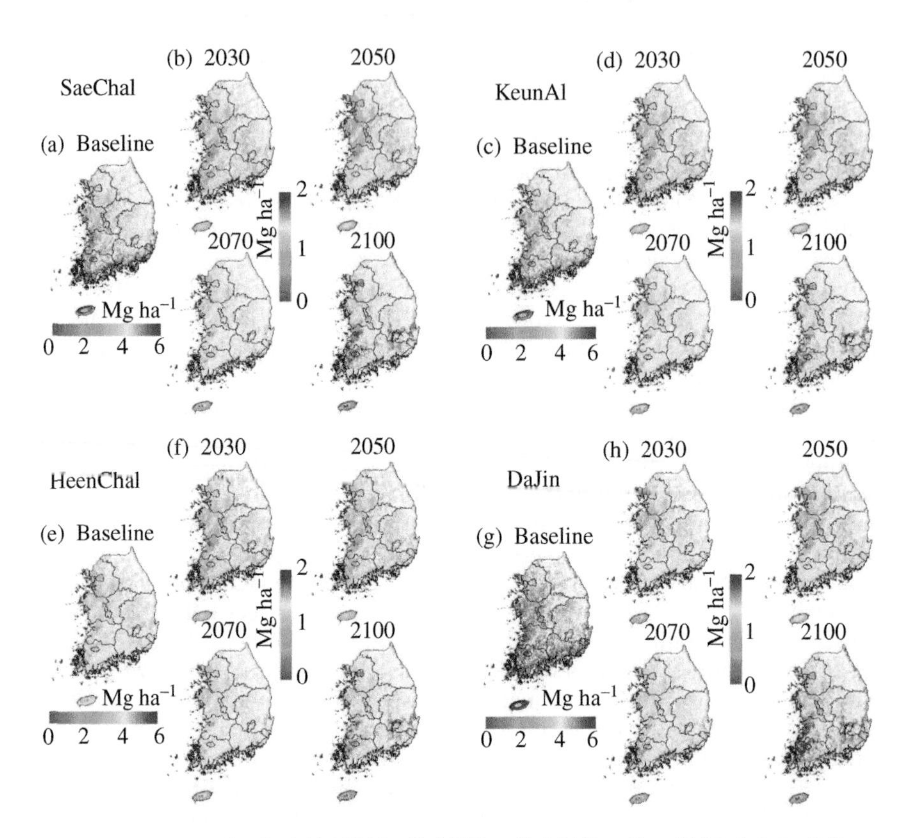

Figure 7.6 Projections of grain yields of (a) SaeChal, (c) HeenChal, (e) KeunAl, and (g) DaJin in South Korea for the current year (baseline), and changes in projected grain yields for future years (2030, 2050, 2070, and 2100) under RCP 8.5 (b, d, f, and h) scenarios in relation to the baseline, modified from Ko et al. (2019).

et al., 2009). There have been some efforts to simulate geospatial variations in crop productivity based on geospatially variable climate in determination of impacts of climate change on staple crops, such as paddy rice in Southeast Asia (Chun et al., 2016; Li et al., 2017) and maize and bean in East Africa (Thornton et al., 2009). To the best of our knowledge, the current research is the first to simulate the effects of climate change on barley with a fine grid (3 km)-based regional yield projection by categorizing regional productivity using *k* means clustering.

Management Options and Outlines of the Geospatial Crop Projections under Climate Change as a Tool to Guide Management by Producers

A study was conducted to simulate the cultivars under the combined effects of temperature, precipitation, solar radiation, and CO_2 climate change (CC) scenarios to identify the optimal planting window that would ameliorate the negative effects

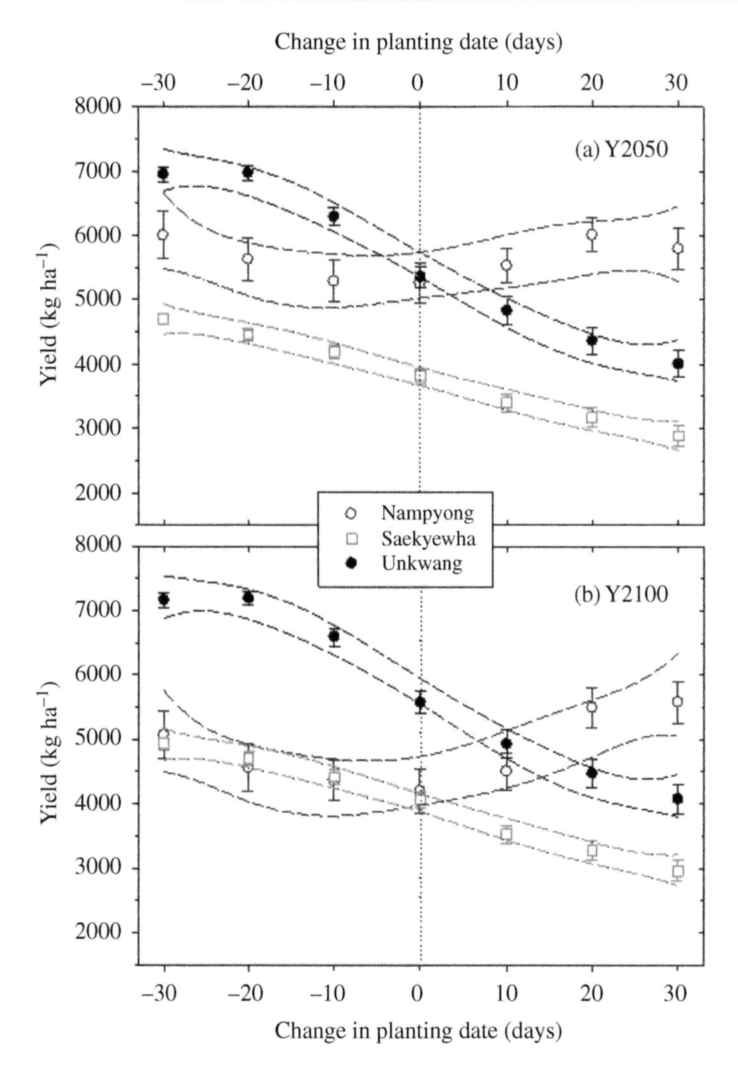

Figure 7.7 Grain yields of three rice cultivars as a function of early and late planting regimes, i.e., early and late plantings vs. the current planting date (0) for the projected years centered on (a) 2050 and (b) 2100 (Kim et al., 2013). Dotted lines and vertical bars represent the base planting date and ±1 SE (*n* = 25), respectively.

of CC on paddy rice (Figure 7.7). These simulations were performed with the cultivars of Nampyong, Saekyewha, and Unkwang planted at 30, 20, and 10 d before and after their actual planting dates to assess whether these simulated planting dates would increase crop yields and compensate for the negative effect of temperature increases in the CC scenarios under investigation. Yields as a function of planting scenarios showed none linear but different curvilinear patterns for all three cultivars in both CC scenarios of 2050 and 2100. Yields reached peaks at

Day 20 early planting for Unkwang, which is an early season cultivar, and at Day 30 early planting for Saegyewha, which is a mid-season cultivar, both in 2050 and 2100. Yields for Nampyeong, a late-season cultivar, reached peaks at Day 20 late planting and at Day 30 early planting for both 2050 and 2100. Dhungana et al. (2006) claimed that future crop production could be adapted to climate change by implementing alternative management practices and developing new crop genotypes that are pre-adapted to projected future climate conditions. Therefore, it appears that early planting of the cultivars may facilitate completion of the plant life cycle before the onset of temperature elevation in the summer. Selecting the appropriate cultivar to adapt to warming conditions is feasible only with sufficient plasticity in the photoperiod and vernalization requirements of crop plants (Masle et al., 1989). The crop model used in the simulations takes into account such adaptabilities.

Results of an ANOVA for simulated barley yields categorized using four factors (RCP scenario, year, region, and cultivars) indicated that year and area were the most important factors, whereas varieties and climate scenario were far less important (data not shown). The scenario–year interaction had the most substantial influence on the total variation in crop yield, followed by year and region. Cultivar and cultivar-related interaction factors were far less important. Therefore, RCP scenario, year, area, and their interactions are important factors in explaining crop yield variations. This fact is coherent with the results shown in Figure 7.8, where crop yield trajectory patterns do not seem to differ significantly between cultivars. The impact of climate change on crop production varied depending on diverse climate types and soil textures. South Korea embraces six climate types: Cfa, Cwa, Dfa, Dfb, Dwa, and Dwb, based on the Koppen–Geiger grouping system (Kottek et al., 2006). Allowing for eight different dominant soil textures spread out in croplands of South Korea based on the USDA classification, we assumed a total of 6×8 groupings of climate–soil types.

The entire country was classified into six geographical units (types) based on *k* means clustering using eight variables (i.e., changes of barley yields from the baseline to 2030, from 2030 to 2050, from 2050 to 2070, and from 2070 to 2100 under both the RCP 4.5 and RCP 8.5 scenarios). The distribution of each of these six landscape types is shown in Figure 7.9. Contour plots (Figure 7.9a,g,m,s) show the distribution of geographical unit Type 1 in the country. Type 1 is scattered over South Korea. Type 1 exists even in regions in which other units dominate. On Jeju Island, Type 1 is concentrated in central, mountainous areas. Type 2 is focused in the southwestern part of South Korea, particularly in Chonnam Province. There are substructures even in Chonnam. Chonnam has a mixture of Types 1 and 2. The coastal Jeju region is a unique geographical unit of Type 3. Both Types 4 and 6 stretch from northwestern to southeastern South Korea. However, Type 6 shows a broader distribution than Type 4. Chungnam mainly contains Type 6. Type 5 is distributed in coastal areas in the south, excluding Jeju Island. The projected impact of climate change on different geographical units is shown in Figure 7.9. Mean grain yields of all barley cultivars in the future under the RCP 8.5 and RCP 4.5 scenarios compared with those at the baseline are likely to vary for different geographical types. Mean grain yields under RCP 8.5 were likely to increase gradually in Type 1 and rapidly in Type 2 as time progressed. However, those under RCP 4.5 are likely to maintain small

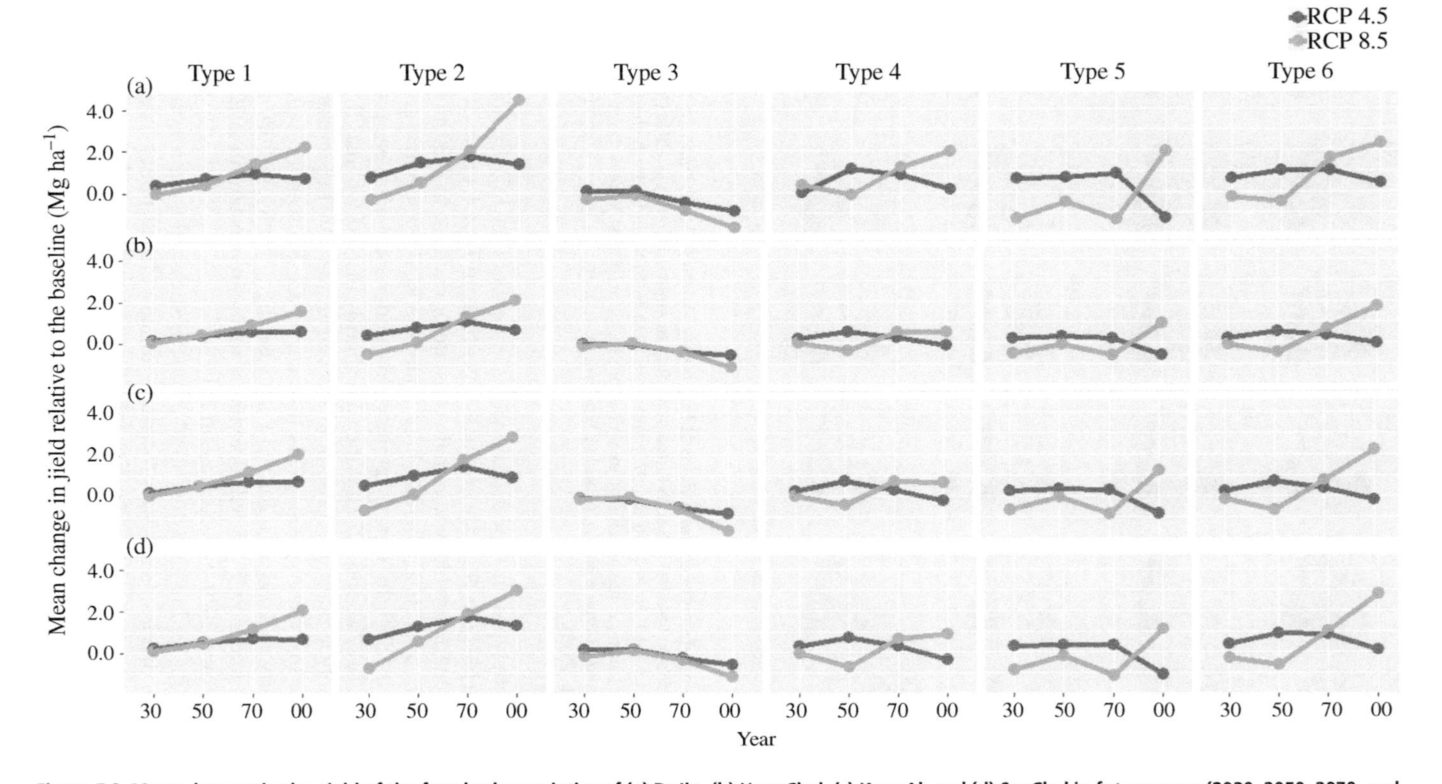

Figure 7.8 Mean changes in the yield of the four barley varieties of (a) DaJin, (b) HeenChal, (c) KeunAl, and (d) SaeChal in future years (2030, 2050, 2070, and 2100) in relation to the baseline of six *k* means clusters (see Figure 7.8) of South Korea (Ko et al., 2019).

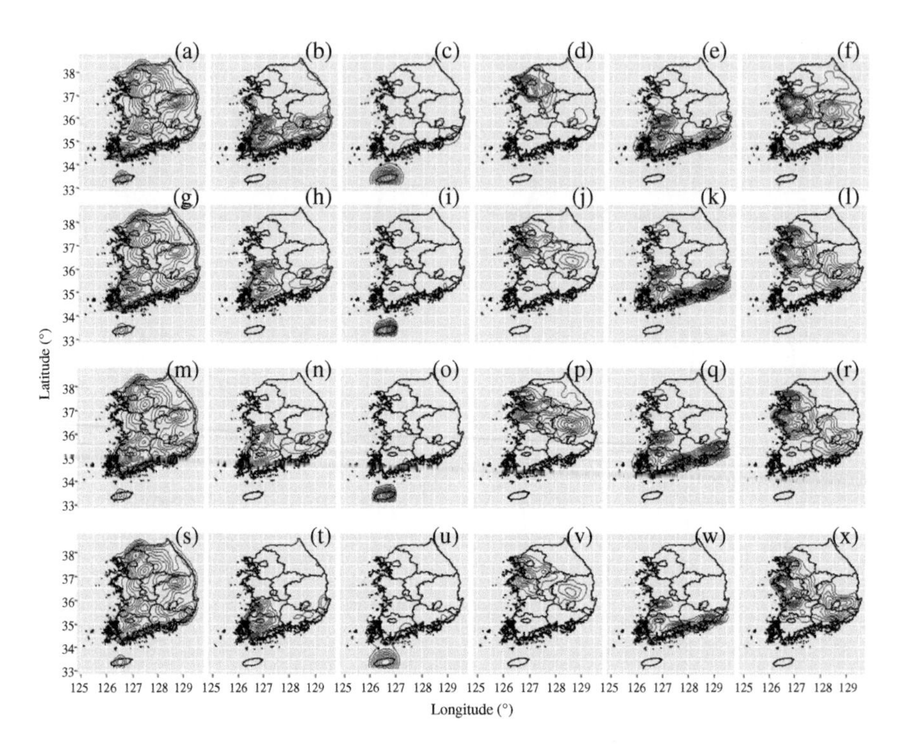

Figure 7.9 Grouping of simulated grain yields of four barley varieties of (a–f) DaJin, (g–l) HeenChal, (m–r) KeunAl, and (s–x) SaeChal in South Korea according to eight different variables (from now to 2030, from 2030 to 2050, from 2050 to 2070, and from 2070 to 2100 under both the RCP 4.5 and RCP 8.5 scenarios) using *k* means cluster examination: (a, g, m, and s) Type = 1; (b, h, n, and t) Type = 2; (c, j, o, and u) Type = 3; (d, j, p, and v) Type = 4; (e, k, q, and w) Type = 5; and (f, l, r, and x) Type = 6 (Ko et al., 2019).

increases in both types. In Type 3, grain yields under both the RCP 8.5 and RCP 4.5 scenarios were likely to decrease gradually over time. In Type 4, grain yields under the RCP 8.5 scenario are likely to remain constant or decrease in 2030 and 2050 but increase in 2070 and 2100, whereas those under the RCP 4.5 scenario are likely to increase only in 2050. In Type 5, grain yields under RCP 8.5 were projected to decrease in 2030, 2050, and 2070 but increase sharply in 2100, whereas those under RCP 4.5 are likely to maintain constant in 2030, 2050, and 2070 but decrease significantly in 2100. In Type 6, grain yields under the RCP 8.5 scenario will possibly decrease slightly in 2030 and 2050 but increase significantly in 2070 and 2100, whereas those under RCP 4.5 will potentially maintain constant in 2030, increase slightly in 2050 and 2070, and decrease in 2010.

Clustering allowed us to reduce local characteristics of the entire country from 48 to 6 classes. Decision-making procedures in agricultural policy issues can be country-wide or area-specific. The dissemination of Type 1 might be useful to evaluate the prospect of introducing country-wide climate change policies (see Figure 7.9). The spread of other types might be beneficial to determine area-specific strategies. However, further efforts should be made to obtain detailed classification information on the distribution of each geographical type and relationships of the data with other factors, such as climate variables and geographic characteristics. Limitations of the current study included climate projection models used and the baseline applied. An ensemble approach using multiple climate projection models would better represent the variability in projected climate changes for simulation of the impact of climate change on crops (e.g., Ko et al., 2012).

Summary and Conclusion

The CERES-Barley and CERES-Rice models demonstrated a potential to be utilized at a local to regional scale to investigate the effects of climate change on barley and rice production in the climate of South Korea. We found that barley productivity in the Republic of Korea (ROK) is likely to increase due to CO_2 fertilization effects of enhanced CO_2 concentration on photosynthesis of rice and barley (C3 crops), except for in a few areas, such as Jeju. Meanwhile, CO_2 fertilization effects on rice could either be negated by the yield reductions due to temperature increase, which could be mitigated by appropriate cultivar selection and selecting appropriate time for planting (planting window). We also found that rice production could increase at higher latitudinal and altitudinal regions due to enhanced temperature effects.

The developed GCSM system using the CERES-Barley model in DSSAT was capable of simulating crop productivity under different climate change scenarios. Geographical regions of the country were classified into six groups based on *k* means clustering and the impact of climate change on barley yield. These results demonstrated the likely influence of local climate change on barley grain yields and allowed us to develop counteractive actions for mitigating the adverse impacts in agricultural production. Although the GCSM system developed in this study needs additional development to meet requirements as an independent tool for scientists and stakeholders, we believe that this system could be efficiently used to simulate topographical dissimilarities of the impacts of climate change on crops and to investigate probable solutions for reducing the uncertainty on food security.

Acknowledgments

This research was supported by a Cooperative Research Programs for Agriculture Science & Technology Development (Project no. PJ01010702) funded by Rural Development Administration, Republic of Korea. Partial support was made by the Basic Science Research Program of the National Research Foundation of Korea (NRF), which is funded by the Ministry of Education, Science, and Technology (NRF-2018R1D1A1B07042925).

Abbreviations

ACO_2, ambient carbon dioxide; BL, baseline; CC, climate change; CDF, cumulative distribution function; CERES, Crop Environment Resource Synthesis; CORDEX, coordinated regional climate downscaling experiment; DSSAT, Decision Support System for Agrotechnology Transfer; ECO_2, elevated carbon dioxide; EPIC, Environmental Policy Integrated Climate; FACE, free air CO_2 enrichment; GCMs, general circulation models; GCSM, geospatial crop simulation modeling; GHGs, greenhouse gases; IPCC, Intergovernmental Panel on

Climate Change; RCP, representative concentration pathway; TFRCD, Task Force for Regional Climate Downscaling; TGFC, temperature gradient field chamber; WCRP, World Climate Research Programme; WOFOST, World Food Studies; WRF, Weather Research and Forecasting model.

References

Adams, R. M., Rosenzweig, C., Peart, R. M., Ritchie, J. T., McCarl, B. A., Glyer, J. D., Curry, R. B., Jones, J. W., Boote, K. J., & Allen, L. H., Jr. (1990). Global climate change and US agriculture. *Nature, 345*, 219–224.

Ahn, J., Hur, J., & Shim, K. (2010). A simulation of agro-climate index over the Korean Peninsula using dynamical downscaling with a numerical weather prediction model. (In Korean with English abstract.). *Korean Journal of Agricultural and Forest Meteorology, 12*, 1–10.

Ahuja, L. R., Rojas, K. W., Hanson, J. D., Shafer, M. J., & Ma, L. (Eds.) (2000). *Root zone water quality model. Modeling management effects on water quality and crop production*. Water Resources Publications, LLC.

Asseng, S., Jamieson, P. D., Kimball, B., Pinter, P., Sayre, K., Bowden, J. W., & Howden, S. M. (2004). Simulated wheat growth affected by rising temperature, increased water deficit and elevated atmospheric CO_2. *Field Crops Research, 85*, 85–102.

Baker, J. T., Boote, K. J., & Allen, L. H. (1995). Potential climate change effects on rice: Carbon dioxide and temperature. In C. Rosenzweig, L. H. Allen, Jr., L. A. Harper, S. E. Hollinger, & J. W. Jones (Eds.), *Climate change and agriculture: Analysis of potential international impacts* (pp. 31–47). ASA, CSSA, and SSSA. https://doi.org/10.2134/asaspecpub59.c2

Boote, K. J., Jones, J. W., & Hoogenboom, G. (1998). Simulation of crop growth: CROPGRO model. In R. M. Peart & R. B. Curry (Eds.), *Agricultural systems modeling and simulation* (pp. 651–692). Marcel Dekker.

Chun, J., Li, S., Wang, Q., Lee, W., Lee, E., Horstmann, N., Park, H., Veasna, T., Vanndy, L., Pros, K., & Vang, S. (2016). Assessing rice productivity and adaptation strategies for Southeast Asia under climate change through multi-scale crop modeling. *Agricultural Systems, 143*, 14–21.

Dhungana, P., Eskridge, K. M., Weiss, A., & Baenziger, P. S. (2006). Designing crop technology for a future climate: An example using response surface methodology and the CERES-Wheat model. *Agricultural Systems, 87*, 63–79.

Fu, T., Ha, B., & Ko, J. (2016a). Simulation of CO_2 enrichment and climate change impacts on soybean production. *International Agrophysics, 30*, 25–37. https://doi.org/10.1515/intag-2015-0069

Fu, T., Ko, J., Wall, G. W., Pinter, P. J., Kimball, B. A., Ottman, M. J., & Kim, H.-Y. (2016b). Simulation of climate change impacts on grain sorghum production grown under free air CO_2 enrichment. *International Agrophysics, 30*, 311–322. https://doi.org/10.1515/intag-2016-0007

Gornall, J., Betts, R., Burke, E., Clark, R., Camp, J., Willett, K., & Wilshire, A. (2010). Implications of climate change for agricultural productivity in the early twenty-first century. *Philosophical Transactions of the Royal Society B, 365*, 2973–2989. https://doi.org/10.1098/rstb.2010.0158

Griffin, T. S., Johnson, B. S., & Ritchie, J. T. (1992). *SUBSTOR-potato version 2.0: A simulation model for potato growth and development*. Michigan State University.

Hatfield, J., Boot, K., Kimball, B., Ziska, L., Izaurralde, R., Ort, D., Thomson, A., & Wolfe, D. (2011). Climate impacts on agriculture: Implications for crop production. *Agronomy Journal, 103*, 351–370. https://doi.org/10.2134/agronj2010.0303

Hijmans, R. J., Guiking-Lens, I. M., & van Diepen, C. A. (1994). *WOFOST 6.0: User's guide for the WOFOST 6.0 crop growth simulation model*. Technical Document 12. DLO Winand Staring Centre.

Horie, T., Kropff, M. J., Centeno, H. G., Nakagawa, H., Nakano, J., Kim, H. Y., & Ohnishi, M. (1995). Effect of anticipated change in global environment on rice yields in Japan. In S. Peng, K. T. Ingram, H. U. Neue, & L. H. Ziska (Eds.), *Climate change and rice* (pp. 291–302). Springer–Verlag.

IBSNAT. (1989). *Decision Support System for Agrotechnology Transfer V. 2.10 (DSSAT V. 2.10)*. Department of Agronomy and Soil Science, University of Hawaii.

IPCC. (2007). Summary for policymakers. In B. Metz, O. R. Davidson, P. R. Bosch, R. Dave, & L. A. Meyer (Eds.), *Climate change 2007: Mitigation. Contribution of working group III to the fourth assessment report of the intergovernmental panel on climate change*. Cambridge University Press.

IPCC. (2013). Climate change 2013: The physical science basis. Contribution of working group I to the fifth assessment report of the intergovernmental panel on climate change. In T. F. Stocker, D. Qin, G.-K. Plattner, M. M. B. Tignor, S. K. Allen, J. Boschung, A. Nauels, Y. Xia, V. Bex, & P. M. Midgley (Eds.), *Observations: Atmosphere and surface* (pp. 159–254). Cambridge University Press.

Jones, J. W., Hoogenboom, G., Porter, C. H., Boote, K. J., Batchelor, W. D., Hunt, L. A., Wilkens, P. W., Singh, U., Gijsman, A. J., & Ritchie, J. T. (2003). The DSSAT cropping system model. *European Journal of Agronomy, 18*, 235–265.

Kim, H., Ko, J., Jeong, S., Kim, J., & Lee, B. (2017). Geospatial delineation of South Korea for adjusted barley cultivation under changing climate. *Journal of Crop Science and Biotechnology, 20*, 417–427. https://doi.org/10.1007/s12892-017-0131-0

Kim, H., Ko, J., Kang, S., & Tenhunen, J. (2013). Impacts of climate change on paddy rice yield in a temperate climate. *Global Change Biology, 19*, 548–562. https://doi.org/10.1111/gcb.12047

Kim, H., Lim, S., Kwak, J., Lee, D., Lee, S., Ro, H., & Choi, W. (2011). Dry matter and nitrogen accumulation and partitioning in rice (*Oryza sativa* L.) exposed to experimental warming with elevated CO_2. *Plant and Soil, 32*, 59–71. https://doi.org/10.1007/s11104-010-0665-y

Kim, J., Park, J., Hyun, S., Fleisher, D. H., & Kim, K. S. (2020). Development of an automated gridded crop growth simulation support system for distributed computing with virtual machines. *Computers and Electronics in Agriculture, 169*, 105196.

Kim, J., Sang, W., Shin, P., Cho, H., Seo, M., Yoo, B., & Kim, K. S. (2015). Evaluation of regional climate scenario data for impact assessment of climate change on rice productivity in Korea. *Journal of Crop Science and Biotechnology, 18*, 257–264.

Kim, Y.-U., & Lee, B.-W. (2019). Differential mechanisms of potato yield loss induced by high day and night temperatures during tuber initiation and bulking: Photosynthesis and tuber growth. *Frontiers in Plant Science, 10*, 1–9.

Kirschbaum, M. U. F. (2000). Forest growth and species distribution in a changing climate. *Tree Physiology, 20*, 309–322. https://doi.org/10.1093/treephys/20.5-6.309

Ko, J., Ahuja, L., Kimball, R., Anapalli, S., Ma, L., Green, T. R., Ruane, A. C., Wall, G., Pinter, P., & Bader, D. A. (2010). Simulation of free CO_2 enriched wheat growth and interactions with water, nitrogen, and temperature. *Agricultural and Forest Meteorology, 150*, 1331–1346.

Ko, J., Ahuja, L. R., Saseendran, S. A., Green, T. R., Ma, L., Nielsen, D. C., & Walthall, C. L. (2012). Climate change impacts on dryland cropping systems in the Central Great Plants, USA. *Climatic Change, 111*, 445–472.

Ko, J., Kim, H., Jeong, S., An, J., Choi, G., Kang, S., & Tenhunen, J. (2014). Potential impacts of climate change on paddy rice yield in mountainous highland terrains. *Journal of Crop Science and Biotechnology, 17*, 117–126.

Ko, J., Ng, C. T., Jeong, S., Kim, J., Lee, B., & Kim, H. (2019). Impacts of regional climate change on barley yield and its geographical variations in South Korea. *International Agrophysics, 33*, 81–96. https://doi.org/10.31545/intagr/104398

Kottek, M., Grieser, J., Beck, C., Rudolf, B., & Rubel, R. (2006). World map of the Koppen–Geiger climate classification updated. *Meteorologische Zeitschrift, 15*, 259–263.

Lee, J., De Gryze, S., & Six, J. (2011). Effect of climate change on field crop production in California's Central Valley. *Climatic Change, 109*, S335–S353.

Li, S., Wang, Q., & Chun, J. (2017). Impact assessment of climate change on rice productivity in the Indochinese Peninsula using a regional-scale crop model. *International Journal of Climatology, 34*, 1147–1160. https://doi.org/10.1002/joc.5072

Lobell, D. B., & Field, C. B. (2007). Global scale climate-crop yield relationships and the impacts of recent warming. *Environmental Research Letters, 2*, 1–7. https://doi.org/10.1088/1748-9326/2/1/014002

Lobell, D. B., Schlenker, W., & Costa-Roberts, J. (2011). Climate trends and global crop production since 1980. *Science, 333*, 616–620.

Long, S. P., Ainsworth, E. A., Leakey, A. D. B., Nosberger, J., & Ort, D. R. (2006). Food for thought: Lower-than-expected crop yield stimulation with rising CO_2 concentrations. *Science, 312*, 1918–1921.

Masle, J., Doussinault, G., & Sun, B. (1989). Responses of wheat genotypes to temperature and photoperiod in natural conditions. *Crop Science, 29*, 712–721. https://doi.org/10.2135/cropsci1989.0011183X002900030036x

McMaster, G. S. (1993). Another wheat (*Triticum* spp.) model? Progress and applications in crop modeling. *Rivista di Agronomia, 27*, 264–272.

Olesen, J. E., Trnka, M., Kersebaum, K. C., Skjelvåg, A. O., Seguin, B., Peltonen-Sainio, P., Rossi, F., Kozyra, J., & Micale, F. (2011). Impacts and adaptation of European crop production systems to climate change. *European Journal of Agronomy, 34*, 96–112.

Ramankutty, N., Foley, J. A., Norman, J., & McSweeney, K. (2002). The global distribution of cultivable lands: Current patterns and sensitivity to possible climate change. *Global Ecology and Biogeography, 11*, 377–392. https://doi.org/10.1046/j.1466-822x.2002.00294.x

Rosenzweig, C., Elliott, J., Deryngd, D., Ruane, A. C., Müllere, C., Arneth, A., Boote, K. J., Folberth, C., Glotter, M., Khabarov, N., Neumann, K., Piontek, F., Pugh, T. A. M., Schmid, E., Stehfest, E., Yang, H., & Jones, J. W. (2014). Assessing agricultural risks of climate change in the 21st century in a global gridded crop model intercomparison. *Proceedings of the National Academy of Sciences of the USA, 111*, 3268–3273. https://doi.org/10.1073/pnas.1222463110

Saseendran, S. A., Singh, K. K., Rathore, L. S., Singh, S. V., & Sinha, S. K. (2000). Effects of climate change on rice production in the tropical humid climate of Kerala, India. *Climatic Change, 44*, 495–514.

Thornton, P. K., Jones, P. G., Alagarswamy, G., & Andersen, J. (2009). Spatial variation of crop yield response to climate change in East Africa. *Global Environmental Change, 19*, 54–65. https://doi.org/10.1016/j.gloenvcha.2008.08.005

Tubiello, F. N., Rosenzweig, C., Kimball, B. A., Pinter, P. J., Jr., Wall, G. W., Hunsaker, D. J., LaMorte, R. L., & Garcia, R. L. (1999). Testing CERES-Wheat with free-air carbon dioxide enrichment (FACE) experiment data: CO_2 and water interactions. *Agronomy Journal, 91*, 247–255. https://doi.org/10.2134/agronj1999.00021962009100020012x

Williams, J. R., Jones, C. A., Kiniry, J. R., & Spanel, D. A. (1989). The EPIC crop growth model. *Transactions of the American Society of Agricultural Engineers, 32*, 497–511. https://doi.org/10.13031/2013.31032

8

Abdullah A. Jaradat and
Dennis J. Timlin

Constraints to Productivity of Subsistence Dryland Agroecosystems in the Fertile Crescent: Simulation and Statistical Modeling

Abstract

Subsistence farming in the Fertile Crescent functioned as a quasi-sustainable buffering strategy to counterbalance climatic instability, and may have slowed down, albeit for a while, the inevitable depletion of natural resources and the gradual decline of agroecosystem services. In retrospect, adaptive subsistence farmers in the Fertile Crescent of West Asia seemed to have managed the unavoidable and avoided the unmanageable in response to climate change. An integrated approach, combining simulation modeling and statistical approaches was employed in studying features of subsistence farming in the Fertile Crescent. Annualized yields of traditional and alternative crop rotations derived from field experiments and from archived records for all countries in the Fertile Crescent were used in a simulation and statistical modeling study. The objectives were to identify constraints to sustainable crop production of subsistence farmers and predict yield gaps due to projected climate changes. Spatial (longitude and latitude coordinates of 28 locations in the Fertile Crescent) and temporal variation (the years 2010, 2050, and 2100, respectively, corresponding with RCP2.6, RCP4.5, and RCP8.5 representative concentration pathways, RCP) in soils, maximum temperatures, rainfall amounts, and monthly distribution accounted for major differences in projected annualized crop rotation yields. Consequently, the growing season will be shortened, thus widen-

Enhancing Agricultural Research and Precision Management for Subsistence Farming by Integrating System Models with Experiments, First Edition. Edited by Dennis J. Timlin and Saseendran S. Anapalli.

DOI: 10.1002/9780891183891.ch08

ing current yield gaps between actual and potential crop rotation yields and forcing significant changes in the farming systems and the choice of crops that can be grown by subsistence farmers. With 50% probability, subsistence farmers may be able to achieve as large as 67.0% (under RCP4.5) or as small as 34.0% (under RCP8.5) of their water-limited crop yields; however, whether these yield gaps can be closed agronomically, ecologically, and sustainably remains an open question. Spatiotemporal variation in heatmaps of annualized crop rotation yields, although large and dynamic, did not respond linearly to changes in RCPs, suggesting that complex interactions are behind this response. Although the impact of simulated climate change on subsistence farming was widespread over the Fertile Crescent, it is expected to be more pronounced in the central and northern parts of the Levant, a region most impacted by the Mediterranean climate. As a follow-up to this study, strategies will be developed in silico that will target each of the 28 locations in the Fertile Crescent where yield of subsistence crop rotations are well below their current water-limited yield.

Introduction

The Fertile Crescent (FC) occupies a central location within the larger Middle East region (Figure 8.1a). The large and diverse region of the FC encompasses seven countries (Figure 8.1) that share general similarities in physiography, topography, soil, and weather, but exhibit large within- and among-country differences in these and other natural endowments (Hole, 2009; Jaradat, 2017; Kitoh et al., 2008). The complex relationship between human–environment–natural and managed agroecosystems that has been playing out in the FC of the Middle East for more than 6,000 yr is acknowledged largely as an outcome of the interplay between the climate, land-use change and the plow (Hole, 2009; Jaradat, 2017). The FC encompasses mountain highlands and seacoasts, river valleys, alluvial plains, oases and steppes, large areas suitable for dryland agriculture, as well as arid zones requiring irrigation (Abi Saab et al., 2015; Salah et al., 2018). It is a large and diverse region that spans two geographical zones: the Mediterranean coast of the Levant, and the floodplains of the Tigris and Euphrates (i.e., Mesopotamia). These two regions are "fertile" in the context of the generally arid Middle East, where the Levant gets a steady stream of rain from the Mediterranean,

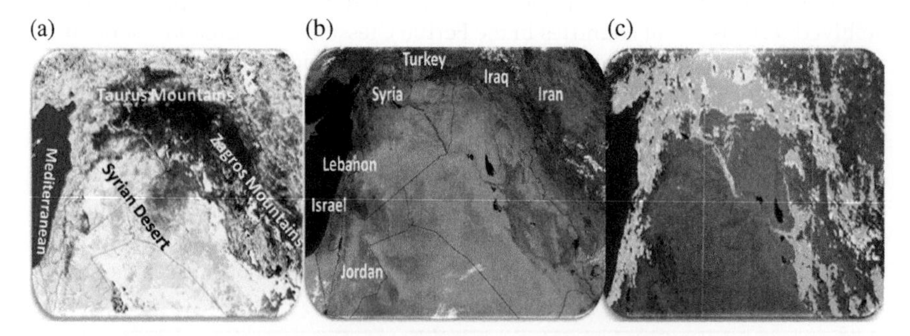

Figure 8.1 Maps of (a) the Fertile Crescent showing physiography, (b) countries and extent of agricultural area, and (c) cropland cover during 2018 growing season. (a and b, Google maps, https://maps.google.com; and c, Google Earth Engine, https://earthengine.google.com, 2018).

enough to support dryland agriculture; while in prehistory, the annual floods of the Tigris–Euphrates used to deposit fertile sediments and could also be used for irrigation (Riehl, 2016; Yatagai, 2015). However, in relation to the larger Middle East, the FC is relatively small. A steep precipitation gradient runs from the wetter north (ca. 400 mm yr^{-1}) to the desert in the south (<150 mm), and from the Mediterranean coast to the interior, so that the band of productive rain-fed agricultural land is relatively narrow.

Most ancient and current settlements in the large plains of the FC were supported by dryland agricultural production, which depended upon soil quality and available rainfall (Braemer et al., 2010). The largest dryland area was occupied by the classic "landscape of Tels" (Figure 8.2), which flourished during the Chalcolithic and Bronze Ages; it continued to support the long-term "staple economy" of numerous Tel-based communities to the present (Alpert et al., 2014; Kitoh et al., 2008). The large land area of Tel-based farming compensated for the relatively lower yields when compared with the more productive irrigated plains, especially in lower Mesopotamia. However, the intensification of such settlements and the conversion of upland areas of the FC to intensive agricultural production created the necessary preconditions for severe soil erosion (Lal, 2013; Pretty & Bharucha, 2014).

The Mediterranean climate regime shaped both the soil formation and the weather of the FC; it will continue to characterize the FC as a "borderland between the Syrian desert in the south, the Taurus mountains in the north, and the Zagros mountains in the east—a cultivable frontier at the edge of the semi-desert" (Alpert et al., 2014; Jaradat, 2017). This climate regime, with its fluctuating magnitudes of annual precipitation and seasonal temperatures, has prevailed for the last ~17,000 yr;

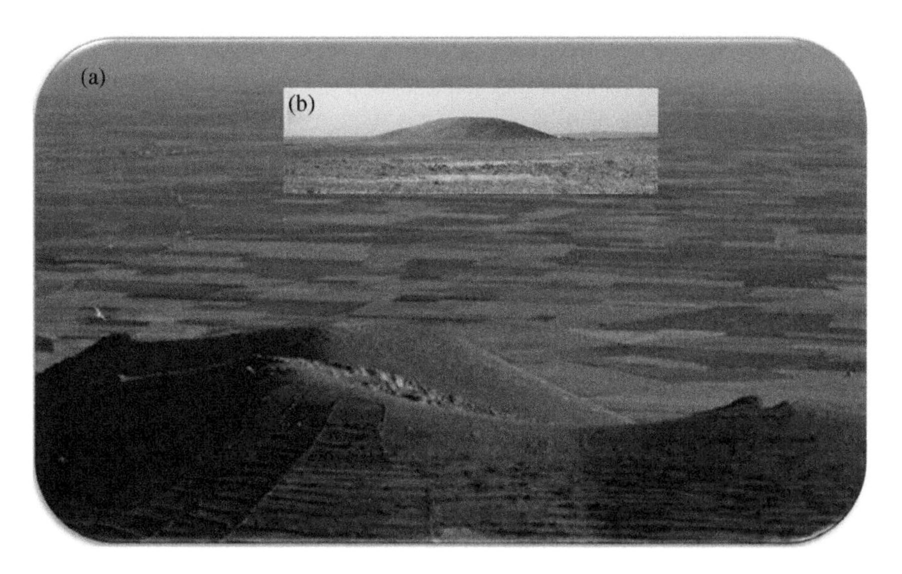

Figure 8.2 Examples of (a) current "Tel" farming system widespread in the Levant and northern Fertile Crescent and (b) climate change/desertification impact on an ancient location in northern Syria (Google Maps, https://maps.google.com, 2018).

however, around 10,000 yr ago, global climate change, rapid increases in CO_2, unique geography, and a responsive human population largely dictated the timing and location of the first agricultural revolution in this part of the Old World (i.e., the Neolithic Revolution) (Hole, 2009; Jaradat, 2017). Climate change and water availability in the FC have major roles in understanding human adaptive capacities while facing long-term environmental changes. Historical (Alpert et al., 2014; Waha et al., 2017) and more recent socioeconomic and demographic changes in the FC (Hole, 2009) have been attributed, at least in part, to climate change.

Agriculture in the FC is inextricably linked to soil quality and water supply, whether from rainfall or irrigation (Ryan et al., 2008; Sommer et al., 2014; Van Duivenbooden et al., 2000). Soils of the FC are largely variable, predominantly calcareous, and frequently phosphate-deficient with variable depth and texture determining the maximum amount of water that can be stored and hence the effective length of the growing season (Ryan et al., 2008). The FC is predominantly classified as an arid to semi-arid region with cool and wet winters (from October to May; wet season), and hot and dry summers (from June to September; dry season). However, a small part is classified as semi-humid, where rainfall exceeds 600 mm and summer temperatures are below 35 °C. A strong precipitation gradient predominates in the FC from the wet mountain areas located in the north and northeast (Taurus and Zagros Mountains, respectively) reaching >1,000 mm, to a very dry regime in the Syrian desert in the south, where annual precipitation rarely exceeds 200 mm. Rainfall mostly occurs during the wet season, with a maximum in December to February, whereas the summer months are practically dry. Precipitation over the FC is mostly produced from three systems: cyclonic disturbance over the Mediterranean, convective precipitation created via strong convective instability given enough water vapor, and orographic rainfall (Sommer et al., 2014; Tadesse et al., 2017; Van Duivenbooden et al., 2000); in general, the lower the annual rainfall, the larger its intra- and interannual variation (Van Duivenbooden et al., 2000; Waha et al., 2017). This variability may have had large impact on long-term sustainable crop production in the FC.

Rainfall patterns, intensity, and dry spells during winter months (middle of October to late April) affect crop establishment, growth, development, and yield (Bassu et al., 2009; Dixit & Telleria, 2015). Fall-sown (winter-growing) crops must often rely on stored soil moisture when they are growing most rapidly at the onset of higher temperatures in the spring months (van Oort et al., 2017). Therefore, rainfall timing and amount, along with soil types (Kelley et al., 2018; Lagacherie et al., 2018), determine the maximum amount of water that can be stored and hence the effective length of the growing season (Figure 8.3). The magnitude of the change in crop phenology and yield of small grain crops are significantly affected by the length of the growing season, but not by soil types, although, clay soils may cause terminal water deficit when crops are actively growing in the spring (Ghanem et al., 2015; Qadir et al., 2018). The projected changes in rainfall timing, decreasing rainfall intensity, and rising temperatures will lengthen the dry season in the FC and will impact the length of the growing season, therefore forcing changes in farming and cropping systems, or even the crops that can be grown by subsistence farmers.

Large soil variation in the FC resulted from the long-term and cumulative influence of climate, topography, and human impact (e.g., land-use change and management practices). Most studies in the FC showed that both climatic and

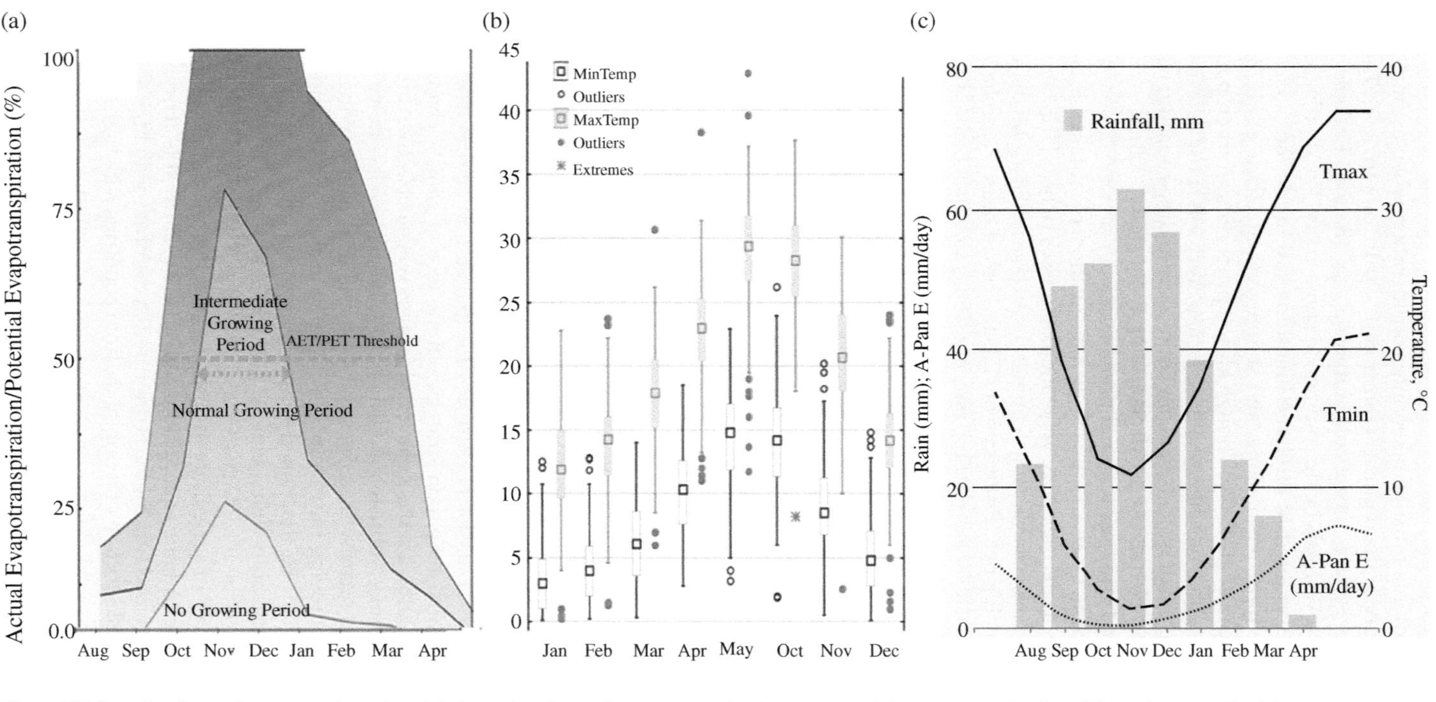

Figure 8.3 Length of growing season based on (a) the ratio of actual evapotranspiration to potential evapotranspiration, (b) maximum and minimum temperatures, and (c) monthly rainfall during a typical growing season in the Fertile Crescent.

edaphic constraints must be taken into consideration when estimating the production capacity of the region (Braemer et al., 2010, for a review). The major soil orders in the FC are Inceptisols, Lithosols, or shallow Entisols, as well as Aridisols, with Vertisols being common in some areas. The dominance of calcareous soils is attributed to the large limestone residuum; these soils are highly variable in texture, depth, slope, and stoniness (Lagacherie et al., 2018; Mrabet et al., 2012; Ryan et al., 2008). Though all soil properties affect crop management and water use efficiency, heavy clay soils (e.g., Vertisols) pose a limitation on tillage operations; shallow soils have limited water-holding capacity; whereas sloping soils are prone to erosion (Mrabet, 2011; Ryan et al., 2008).

The mild and rainy winters, and dry hot summers in the FC were conducive to the establishment, by an adaptive human population, of sustainable cropping systems based on a specific native vegetation of grasses (wheat [*Triticum aestivum* L.] and barley [*Hordeum vulgare* L.]); legumes (lentil [*Lens culinaris* Medikus], chickpea [*Cicer arietinum* L.], pea [*Pisum sativum* L.], and broad bean [*Phaseolus vulgaris* L.]); evergreen (olive [*Olea europaea* L.], pine nut [*Pinus koraiensis* Siebold & Zucc.]) and deciduous (almond [*Prunus dulcis* Miller], fig [*Ficus carica* L.], grape [*Vitis vinifera* L.]) fruit trees; summer and winter vegetables (okra [*Abelmoschus esculentus* (L.) Moench], eggplant [*Solanum melongena* L.], squash [*Cucurbita* spp.], watermelon [*Passiflora laurifolia* L.], cucumber [*Cucumis sativus* L.]); medicinal plants; cultivated herbs; and semi-domesticated plants gathered from natural or semi-managed ecosystems (reviewed in Jaradat, 2017) (Figure 8.1c). Agriculture in the semi-arid FC has been sustained, despite past and more recent changes in climate, through modifications in the balance between agriculture and livestock raising, in expansion and retraction of settlement, and in societal organization of production; however, most recently, this balance has been dramatically disrupted with far-reaching environmental and socioeconomic instabilities (Hole, 2009).

For centuries, a major part of crop production in the FC was mainly for subsistence (Hole, 2009; Jaradat, 2017; Jaradat et al., 2010); however, for most of its history, dryland agriculture in the FC was a subsistence but hardly sustainable option, especially in times of environmental stress (Lal, 2013; Thornton et al., 2018). Its increasingly arid and semi-arid climate poses a formidable challenge for sustainable farming, and to subsistence farmers due to their limited resource base and lack of adaptation and mitigation strategies in the face of climate change (Carranza-Gallego et al., 2018). Unlike commercial farmers whose objectives are to maximize yield and income, subsistence farmers are inclined to be more interested in improving food security by reducing risk of crop failure, improving the return on inputs, and improving the quality of life (Berry et al., 2015). Occasionally, subsistence farming in the FC functioned as a sustainable buffering strategy to counterbalance climatic instability (Johansen et al., 2012; O'Leary et al., 2018; Riehl et al., 2014).

Food imports of the FC (as virtual water) approached ~80% of its needs, whereas the share of agriculture in gross domestic product of most countries in the FC is negligible (Adamo et al., 2018; Adhikari et al., 2018). It is alarming that most of the imported food into the FC is produced in countries with high soil loss potential, and a sizable portion of it originates from countries with high water shortages (Fridman & Kissinger, 2018). However, agriculture continues to play a vital role in livelihood strategies of rural people, in general, and subsistence farmers, in particular (Mrabet et al., 2012; Ryan et al., 2008). Since the 1950s, the subsistence farming systems in the FC, and in other parts of West Asia and North Africa, experienced several interlocking stressors other than climate change and climate variability; these include increased population pressure and associated environmental degradation (Lal, 2013; Sommer et al., 2014), regionalized and globalized markets, and protectionist agricultural policies (Fridman & Kissinger, 2018). Since the 1980s, these socioeconomic, demographic, and geopolitical factors led subsistence farmers to take drastic measures that resulted in land-use change of the agricultural landscape and significant reduction in the production base in several countries of the FC (Figure 8.4).

Smallholder and subsistence farmers, particularly in hilly regions of the FC, practice mixed farming as a means of "sustainable intensification" (Pretty & Bharucha, 2014), to integrate crop–livestock production, diversify production and meet local consumption needs, maintain soil fertility by recycling soil nutrients, and allowing the introduction and use of rotations between various grain crops, forage legumes, vegetables, and fruit trees (Figure 8.5). Farmers in this region, for thousands of years, adapted to climate change by changing their farming patterns and crop rotations through earlier sowing, using shorter duration crops, and switching to crops that are more tolerant to abiotic stresses (Alpert et al., 2014; Hole, 2009; Kelley et al., 2018). These adaptations may have had mitigating effects by sequestering C in the soil. More recently, however, a major concern is the predicted large negative impact of climate change and its concomitant rising CO_2 on subsistence

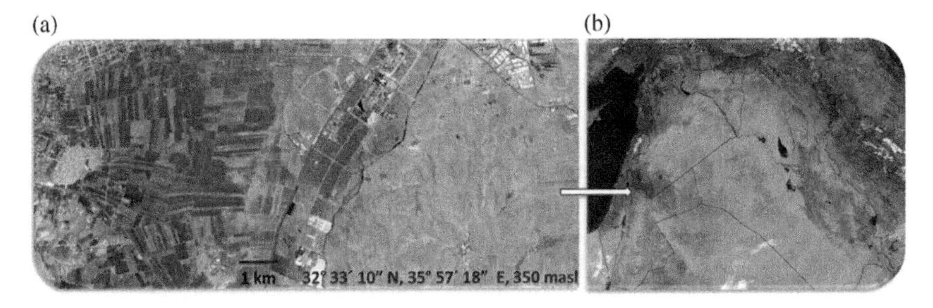

Figure 8.4 Example of land-use change and the sudden transition from semi-arid (a) to arid zone (b) of the Levant in the Fertile Crescent (Google Maps, https://maps.google.com, 2018).

Figure 8.5 Example of (a) mixed farming and (b) traditional management practices in the Fertile Crescent (Google Maps, https://maps.google.com, 2018).

farming throughout the FC. The capacity for CO_2 fertilization to provide increased future food production, and its impact on agroecosystem processes, need to be carefully assessed through further model development and sensitivity testing and analyses. Nevertheless, intensification of subsistence farming will likely push agroecosystems beyond their natural capacity and may result in more environmental degradation and further loss of agroecosystem services (Lancelotti, 2016; Mahon et al., 2017, 2018; Malek & Verburg, 2017; Malek et al., 2018; Pretty & Bharucha, 2014).

Subsistence farmers of the FC region developed over centuries a wide range of cropping systems and crop rotations where water is a major limiting factor to agricultural productivity (Ryan et al., 2008; Salah et al., 2018). Although most crops grown by these farmers are of local origin, the region is chronically food-deficient (Berry et al., 2015). Small grains (i.e., wheat and barley) occupy large areas in the semi-arid and arid parts of the region; however, their yields are well below the world average (Balkovič et al., 2014; Dixit et al., 2018; Pala et al., 2011). Early farmers of the FC, whether by choice or by necessity, realized that yield stability was more valuable for their livelihoods than maximum yield; such choice was supported by heterogeneous natural and managed species they used for food and feed sources. Most likely, nutritional considerations outweighed grain yield as a selection criterion and was reflected in the founder crops repertoire (Jaradat, 2017).

Arid and semi-arid subsistence farms are biophysically and socioeconomically heterogeneous, even over short distances, and are more vulnerable to climate change than those under better climatic conditions (e.g., semi-humid). Recently, extreme events threatened a further decline in the region's rainfall and available water resources, thus leading to a significant change in its climate. Decreasing trends in rainfall have far-reaching environmental, agricultural, and socioeconomic implications on a transitional sensitive region narrowly located between the wet Mediterranean and the arid interior, especially an area located beyond the Levant and well into central and southeastern Syria (Figure 8.1b); this region is subjected to large intra- and inter-annual rainfall variability and its agricultural potential has been steadily declining for a few decades; it culminated in the case of Syrian drought in 2008. In addition to aridity, farming in these areas faces multiple biophysical (e.g., vulnerability to declining rainfall, low soil fertility, and soil erosion) and socio-economic (e.g., small holdings, sub-optimal management, and aging farmers) challenges (Jaradat et al., 2010; Michalscheck et al., 2018). Most smallholder subsistence farmers in parts of the FC and in many developing countries, are resource-poor (Johansen et al., 2012; Riehl et al., 2014; Tittonell & Giller, 2013; Tadesse et al., 2017). Their farming practices typically are based on manual labor or draft animals and seldom rely on energy-based inputs, such as fertilizers, pesticides, and mechanized farm operations. Their farms are usually small (<2 ha), rely on nutrient mining, and are largely rain-fed (Hole, 2009; Jaradat et al., 2010; Lal, 2013). Crop yields in subsistence farming have gradually declined over the years and threatened the livelihood of many people. Nutrient deficiency (Ryan et al., 2012) is a major constraint to sustainable crop production due to the subsistence nature of farming systems in which nutrients are rarely returned to the field (Johansen et al., 2012; Lal, 2013).

Most farming systems in the FC are facing similar problems and challenges, some of which are the products of physiography, changing climate, and a deteriorating production base, whereas others are caused by land-use change due to a growing population pressure on limited land and water resource bases (Cui et al., 2018; Mrabet, 2011; Mutyasira et al., 2018). However, research and development priorities to remedy these problems have been identified, with limited implementation and unsatisfactory results (Michalscheck et al., 2018; Tesfaye & Seifu, 2016; Whitbread et al., 2010).

Past statistical studies and output of integrated assessment models indicated that climate change will impact agricultural yields in the FC (Dixit & Telleria, 2015; Dixit et al., 2018; Tadesse et al., 2017). Shifts in timing, amounts, and patterns of annual rainfall, along with rising temperatures due to climate change, pose a daunting challenge, especially to subsistence farmers (Bassu et al., 2009; Corbeels et al., 2018; Jaradat et al., 2010). Grain yield variability of major and minor crops is large within and among countries of the FC (Benli et al., 2007; Moeller et al., 2007; Pala et al., 2011). For example, average yields of dryland wheat range from 0.9 to 4.5 $Mg\,ha^{-1}$ (Pala et al., 2011); however, most subsistence farmers rarely produce more than 2.0 $Mg\,ha^{-1}$ of wheat although water-limited yields can be as high as 6.0 $Mg\,ha^{-1}$ (Balkovič et al., 2014). Results gleaned from previous simulation studies (e.g., Dixit & Telleria, 2015; Dixit et al., 2018; Jaradat et al., 2010; Sommer et al., 2012) suggested that the amount and timing of rainfall, temperature, soil C, and nutrients highly likely limit productivity and will impact future sustainability in the FC because of climate change (Firbank et al., 2018; Ryan et al., 2012). A crop rotation of a cereal crop (wheat or barley) after a legume crop (lentil or chickpea) or fallow, has its origins in Mediterranean agriculture, the practice is of lasting relevance in dryland agriculture where rainfall and native soil fertility are the limiting factors (Balkovič et al., 2014; Ryan et al., 2008). The more diverse crop rotations, regardless of input levels, were more productive and their baseline and simulated yields were less variable than traditional, short-term rotations, especially when germplasm responsive to external inputs and N-fixing legumes were included in crop rotations (Reckling et al., 2016); additional yields were realized when crop residues were incorporated into the soil (Sommer et al., 2014). Negligible nutrient loss is expected in subsistence farming systems due to limited soil moisture (Zittis et al., 2014); however, soil loss due to erosion was identified as a major factor impacting sustainability (Jaradat et al., 2010; Kelley et al., 2018; Lal, 2013; Ryan et al., 2012). In addition, subsistence agroecosystems may incur lower depletion levels of their meager C stocks in response to diversification, crop–livestock integration, and the inclusion of legume crops in more complex crop rotations of long duration (Lagacherie et al., 2018; Lal, 2013).

Agroecosystem classification based on altitude, rainfall, and growing degree days, delineated five generally overlapping systems where land-use options and management practices are dictated by rainfall patterns and soil quality and productive capacity. Each class can be characterized by its problems, challenges, and research and development priorities (Table 8.1) (van Oort et al., 2017). The large expanses of the arid (grazing) systems (<200 mm of annual rainfall), bordering, and occasionally penetrating the Syrian desert, is not suitable for agricultural production due to low and erratic rainfall and poor soil quality; however, in years of above-normal rainfall, an opportunistic barley crop can be produced, especially where rainfall harvesting is feasible (Saïdi et al., 2018). The arid (grazing) system overlaps with the "marginal" dryland farming system (250–300 mm of annual rainfall), where barley, and occasionally wheat, is the dominant crop; the former is usually integrated with livestock

Table 8.1 Main agroecosystems, land use, management practices, problems, and research and development priorities in the Fertile Crescent in response to climate change

Altitude (m)	Rainfall (mm)	GDD	Land use management	Problems and challenges	R&D priorities
Arid (grazing) systems					
200–500	<200	>4,000	Grazing, opportunistic barley crop	Loss of biodiversity; low fertility; soil erosion	Grazing shrubs; biodiversity rehabilitations, water harvesting
[Marginal] Dryland farming systems					
500–750	250–300	<4000	Barley (or wheat)/fallow Grazing	Low productivity; drought	Forage legume; crop/livestock integration; organic matter; planting date; intercropping; reduced tillage. Rotation: barley–vetch or wheat–vetch
Rain-fed mixed farming systems					
100–1,000	450–650	5,000–7,000	Variable land use systems; small grain and food legume crops, fruit trees; some crop–livestock integration	Decreasing land suitable for cropping	Crop diversification (specialty crops) with supplemental irrigation (water harvesting; protected agriculture); reduced tillage. Rotation: wheat–food legume–summer crop–forage legume–oilseed crop
Cereal–livestock farming systems					
800–1,000	250–450	3,000–4,000	Small grains; small ruminants; smallholder farms; subsistence	Inter- and intra-annual variable/erratic rainfall; drought; degraded land	New land use systems based on larger/consolidated farms; long/complex crop rotations Rotation: wheat–food legume–summer crop; planting date
Mountainous mixed farming systems					
>750	>600	2,500–3,500	Small grains; fruit trees; vegetables; terracing; some supplemental irrigation; smallholder farms; subsistence	Variable rainfall; soil erosion, low fertility.	Watershed management; crop–livestock integration; terracing; perennial crops; recued tillage.

Note. R&D, research and development; GDD, growing degree days.

and grazing. Low productivity, due to rainfall uncertainty and occasional drought, is a characteristic of this system. The rain-fed mixed farming system occupies (geographically, agriculturally, and economically) a central place in the FC. Land area, soil types, and rainfall patterns are most suitable for mixed farming and the cultivation of multiple crops; however, smallholdings and soils prone to water and tillage erosion prevail, especially at high elevations (Tittonell & Giller, 2013). The cereal–livestock system is, spatially and economically, intermediate between the last two, and shares some of their salient features, especially the widespread small, subsistence family farms (Shakoor et al., 2013; Tesfaye & Seifu, 2016). Finally, the mountainous mixed farming system, with its ample precipitation but poor soils and limited land area suitable for cropping, is dotted by small, subsistence farms producing some small grains, fruits, and vegetable crops.

In this study, archived data for each country from FAOSTAT (FAO, 2018b) and simulated data generated from 28 locations in seven countries of the FC were used. This study explores the prospects of subsistence farming in the FC based on the representative concentration pathways (RCPs) developed by the Intergovernmental Panel on Climate Change (IPCC) for its Fifth Assessment Report (IPCC, 2014). The simulations were based on the projections from downscaled global climate models from the CMIP5 (Coupled Model Intercomparison Project) forced by the medium-term (2010–2050) and long-term (2010–2095) Representative Concentration Pathway 4.5 (i.e., RCP4.5 M and L) and Pathway 8.5 (i.e., RCP8.5 M and L), in addition to RCP2.6 greenhouse gas radiative forcing (Jones et al., 2011; Jones & Thornton, 2013). According to the IPCC Fifth Assessment Report, projected changes in temperature (compared with the reference period 1985–2005) for RCP4.5 suggest an increase of about 2.0 °C for most of the FC by the 2050s; however, stronger warming is projected by the end of the 21st century, to an overall increase of 3.0 °C. Projections for RCP8.5 indicated an additional 1.0 °C temperature increase overall, with warming intensifying to an increase of 4.8 °C by the end of the 21st century.

- RCP2.6: Immediate stabilization; sustained net negative CO_2 after 2070; radiative forcing is stabilized at about 2.6 W m^{-2} after 2100. Temperature increase <1.5 °C.; CO_2–eq < 450.
- RCP4.5: Intermediate stabilization; radiative forcing is stabilized at about 4.5 W m^{-2} after 2100. Temperature increase >1.5 °C.; CO_2–eq > 550.
- RCP8.5: High pathway, radiative forcing reached >8.5 W m^{-2} by 2100 and continues to rise. Temperature increase 2.6–4.8 °C.; CO_2–eq. 900 to 1,000 (van Vuuren et al., 2011).

Therefore, rising temperature is expected to remain the most significant feature of global warming in the FC for the next 70–80 yr, if not longer (Rosenzweig et al., 2014; Saymohammadi et al., 2018; Shakoor et al., 2013).

The objectives of the study in the FC were to (a) review the natural endowments and practices of smallholder subsistence farmers, (b) quantify potential yield of traditional and alternative crop rotations across the rainfall gradient and soil types under medium- and long-term climate change scenarios, and (c) assess the constraints to crop rotation productivity using simulation and statistical modeling.

Materials and Methods

Countries and Locations

Twenty-eight locations have been selected for the study and represent the two major geographical zones in the FC, i.e., the Mediterranean and Mesopotamia. These locations represented all major soil types and rainfall patterns, as well as temperature regimes and subsistence agroecosystems in the FC and the larger Middle East Region.

A summary of modules, methods, and procedures used to generate, manage, simulate, and statistically analyze data is presented in Table 8.2. Detailed information is presented under the appropriate headings or sub-headings that follow.

Weather Data

Future climate data for each of the 28 locations were downscaled using MarkSim DSSAT weather file generator. An ensemble of 16 general circulation models (GCMs) with each of three RCPs (i.e., RCP2.6, RCP4.5, and RCP8.5) were constructed and each combination was run for 99 "years." These RCPs represent three different scenarios that stabilize radiative forcing at 2.6, 4.5, and 8.5 W m^{-2}, respectively (Table 8.2) (Riahi et al., 2011; Ruane et al., 2018; Thomson et al., 2011; van Vuuren et al., 2011).

Daily rainfall and temperature output from combinations of 99 location–GCM–RCP and two time period runs (medium-term, 2010–2050; and long-term, 2010–2095) were used to drive a validated version of the Decision Support System for Agrotechnology Transfer (DSSAT) (Hoogenboom et al., 2015; Jones et al., 2011).

Basic "traditional" management practices as described in relevant references for each location (or group of locations with similar agroecological characteristics) were used in simulating a "traditional" package of practices, including crop rotations and minimal inputs, whereas modified packages (Table 8.2) were used to simulate alternative crop rotations (Pala et al., 2011; Ryan et al., 2008).

Sources of Crop Data

Three sources of crop data were used in the study: (a) primary or secondary statistics on biomass and grain yield that were extracted from relevant country- or location-specific publications and, where possible, contrasted with FAOSTAT data (FAO, 2018b); (b) the most recent FAOSTAT database for each crop and country was downloaded and tabulated for statistical analyses; and (c) simulated data based on the simulation protocols as described in Table 8.1.

Crops

Traditional crop species included barley, chickpea, lentil, and vetch (*Vicia sativa* L.), whereas alternative crop species included medic (*Medicago lupulina* L.), safflower (*Carthamus tinctorius* L.), and sorghum [*Sorghum bicolor* (L.) Moench] (Table 8.2). The Generalized Likelihood Uncertainty Estimation (GLUE) program, which was developed using the R statistical programming language, was used to estimate genetic coefficients with 5,000 model runs for each crop in the DSSAT simulation software (He et al., 2010; Jones et al., 2011).

Table 8.2 Summary of methods, inputs, simulation, and statistical procedures used in the current study

Simulation procedure	Software	Module	References
	DSSAT v. 4.7.0	CERES; DSSAT–CROPGRO; DSSAT–CSM; CENTURY–Soil module	Hoogenboom et al. (2017) https://dssat.net
Global circulation models	Average of 16 MarkSims: BCC-CSM1–1, BCC-CSM1–1-M, CSIRO-MK3–6-O, FIO-ESM, GFDL-CM3, GFDL-ESM2G, GFDL-ESM2M, GISS-E2-H, GISS-E2-R, IPSL-CMSA-LR, IPSL-CMSA-MR, MIROC-ESM, MIROC-ESM-CHEM, MIROCS, MRI-CGCM3, and NorESM1-M.		http://ipec-ddc.cru.uea.ac.uk http://gismap.ciat.cgiar.org/MarkSimGCM

Data source	**IPCC**	**Time scale**	**Acronym**	**Description**			http://www.ipcc-data.org
				Current and alternative crop rotations/management			**RCP details**
Land capability class (LCC)				**LCC1**	**LCC2**	**LCC3**	
Representative concentration pathways (RCP)		2018	RCP2.6	(1) Wheat–chickpea/lentil–sorghum–safflower[a]	(4) Wheat–chickpea/lentil (5) Wheat–wheat–vetch/medic[a]	(8) Barley–fallow (9) Barley–fallow–vetch/medic[a]	Immediate stabilization; sustained net negative CO_2 after 2070; radiative forcing is stabilized at about 2.6 W m^{-2} after 2100. Temperature <1.5 °C.; CO_2–eq <450 (van Vuuren et al., 2011).
	2010–2050 (medium-term) 2050–2095 (long-term)		RCP4.5 M RCP4.5 L	(2) Wheat–chickpea/lentil–wheat–fallow[a]	(6) Wheat/barley–vetch/medic–sorghum[a]	(10) Barley–weedy fallow–weedy fallow[a]	Intermediate stabilization; radiative forcing is stabilized at about 4.5 W m^{-2} after 2100. Temperature >1.5 °C.; CO_2–eq >550 (van Vuuren et al., 2011).
	2010–2050 (medium-term) 2050–2095 (long-term)		RCP8.5 M RCP8.5 L	(3) Wheat–vetch/medic–fallow[a]	(7) Wheat/ barley–barley/vetch/medic–sorghum[a]	(11) Barley–weedy fallow–weedy fallow[a]	High pathway, radiative forcing reached >8.5 W m^{-2} by 2100 and continues to rise. Temperature 2.6–4.8 °C.; CO_2–eq 900–1,000 (van Vuuren et al., 2011).
			Tillage	Minimum or no-tillage where applicable			
	Crop residue			80% incorporated			
	Fertilizer			Organic/mineral fertilizer inputs as recommended by extension/location characteristics			
	Supplemental irrigation			To predict potential crop rotation yield where rainfall is limiting			

(Continued)

Table 8.2 (Continued)

Simulation procedure	Software	Module	References
Weather downscaling	**MarkSimGCM**	**Location coordinates (Table 8.3)**	http://gismap.ciat.cgiar.org/MarkSimGCM
Statistical analyses	**Software program**	**Modules/procedures**	**Reference**
	JMP 13.2 Pro.	Hierarchical clustering; variance components; partial least squares regression	SAS Institute (2016)
	R_3.5.1	GLUE package	http://www.R-project.org
	TIBCO Statistica 13.3	Data management	TIBCO Statistica v. 13.3

Statistics	**Availability of primary and secondary statistics**						
	Distribution	**Correlations**	**Whole model validation variance**	**Variance components/ prediction**		**Multivariate distance**	**Country**
				Inx	**Nested**		
Experimental unit							
Response to RCP	x	x	x	x		x	
Rotation yield	x	x	x	x			x
Yield gap	x	x	x		x		x
Soil C	x	x	x		x		

Note. LCC, land capability class; RCP, representative concentration pathway; IPCC, Intergovernmental Panel on Climate Change; GCM, general circulation model.
[a] Alternative crop rotations/management.

Crop Rotations

Three traditional and seven alternative crop rotations (Table 8.2) were simulated using the sequence option in DSSAT v. 4.7.0. A crop rotation was defined as a sequence of crops that is fixed (each crop follows a pre-defined order), cyclical (in that it repeats itself), and has a fixed length, whereas the cropping system comprises the rotation, management practices, and inputs (Pala et al., 2011; Reckling et al., 2016; Ryan et al., 2008). Simulated data were extracted from the simulation runs and a relational database was constructed, annualized crop rotation yields were calculated, and then used in subsequent statistical analyses.

Soil Data

Uncertainty in soil data (i.e., physical, biogeochemical, native fertility, erodibility, and so forth) from countries in the FC is being gradually remedied through regional and international efforts, such as Half a degree Additional warming, Prognosis and Projected Impacts Land use scenario experiment (HAPPI-Land; Hirsch et al., 2018), and the Harmonized World Soil Database (HWSD), which represent the most up-to-date world soil map (FAO Soils Portal, FAO, 2018a).

Data on six major soil types (Figure 8.6) were extracted and documented either from cited references (marked with boldface font in the References list) or from the

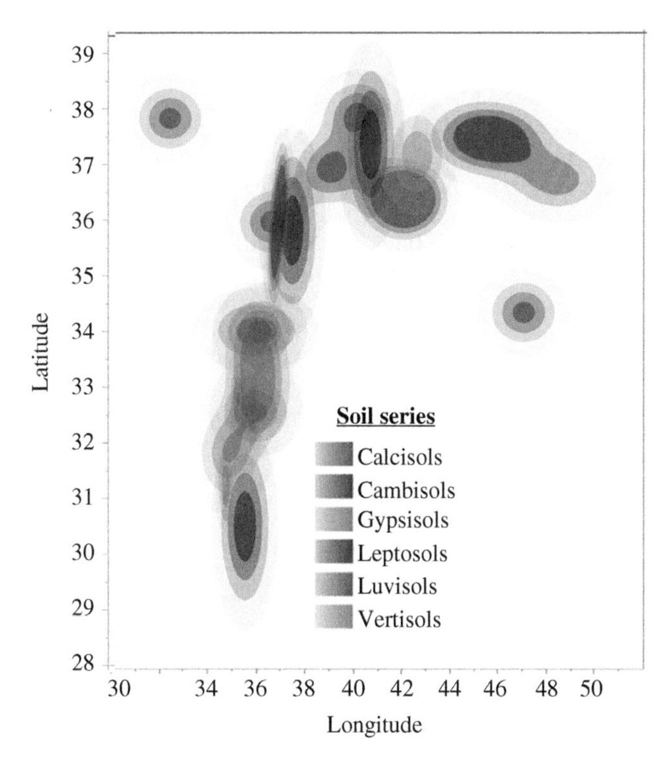

Figure 8.6 Geographical distribution of six major soil series in the rain-fed areas along the longitude and latitude coordinates of the Fertile Crescent.

HWSD, which represent the most up-to-date world soil map (FAO Soils Portal, FAO 2018a). The map incorporates a data table of 48,148 soil profile descriptions related to the various soils associated with each mapping unit at a spatial resolution of 30 arc-seconds (approximately 1 km at the equator).

Simulation Modeling

The Decision Support System for Agrotechnology Transfer (DSSAT v. 4.7.0) includes improved application programs for seasonal, spatial, and sequence (i.e., crop rotation) analyses that assess the economic risks and environmental impacts associated with irrigation, fertilizer and nutrient management, climate variability, climate change, soil C sequestration, and precision management (Hoogenboom et al., 2015; Li et al., 2015). The minimum daily weather data required to run the DSSAT model includes daily solar radiation (MJ m^{-2}), daily maximum and minimum temperatures (°C), and daily precipitation (mm). The model can predict crop yield, resource dynamics such as water, N and C, environmental impact (i.e., N leaching), evapotranspiration, and soil organic matter accumulation (Brilli et al., 2017; Kassie et al., 2016; Nash et al., 2018).

The overall objective of the simulation protocols was to explore the impact of climate change across locations on the annualized biomass and grain yield of traditional and alternative crop rotations. Evaluation of the reported (or measured) data showed that the performance of the DSSAT model was realistic as indicated by the accurate simulation of biomass and grain yield of crop rotations against measured data (Kassie et al., 2016; Li et al., 2015; Silungwe et al., 2018). Due to the large seasonal variability in the Mediterranean environments (Silungwe et al., 2018), a combination of experimental data and simulation results were used to assess crop response to the simulation protocols (Dixit & Telleria, 2015; Önol et al., 2014). Simulation was made under the DSSAT Sequence Analysis program so that the soil nutrient and water dynamics could be continuously transferred from the beginning to the end of the simulation run (Li et al., 2015, 2018). The uncertainty in soil data was taken into consideration during the simulation process (Folberth et al., 2016) by selecting the major soil for simulation runs (>30%) from the various soils associated with each mapping unit (i.e., location) (FAO, 2018a). In addition, simulation results were contrasted with a large number of earlier simulation findings carried out in several parts of the FC at different scales, locations, and times (e.g., Benli et al., 2007; Constantinidou et al., 2016; Dixit & Telleria, 2015; Dixit et al., 2018; Haim et al., 2008; Heng et al., 2007; Moeller et al., 2007, 2014; Önol et al., 2014; Pala et al., 2011; Soltani et al., 2016).

Model Evaluation

Model performance and efficiency in reproducing observed data was assessed using root mean square error (RMSE) and graphical method (Li et al., 2015). Model sensitivity analysis was performed for crop rotation yield using monthly rainfall and percentage rain falling prior to 1 January at two extreme locations and one intermediate location (longitude and latitude coordinates) where comprehensive and long-term data were available. In addition, and where possible, simulated grain yield was contrasted (R^2 estimates) with FAO statistical yield for each country (FAOSTAT, FAO, 2018b).

Statistical Modeling

Observed yield data were estimated as a function of agrometeorological data using partial least squares regression (PLSR) models; results were presented either in equation form or as heat maps relating the estimated crop rotation yield to the independent variables (Asseng, 2013; Lobell & Asseng, 2017; Lobell et al., 2008). The advantages of this approach are its need for minimal data, less computer processing time, and statistical validity of results. However, the information gleaned from statistical modeling is not as comprehensive as the information gleaned from simulation modeling, extrapolation beyond data boundaries is statistically invalid, and agronomic validity or interpretation of results may not be straightforward. Although additional location information can be incorporated in the PLSR models to improve predictability, including location coordinates, soil type, and estimated evapotranspiration derived from the FAO-56 single crop coefficient algorithm, probability of exceedance of both cumulative rainfall and growing degree days, the probability of minimum and maximum daily temperatures exceeding user-defined temperature thresholds, and the probability of heat stress, cold stress, and dry period (i.e., no recorded rainfall) of varying durations; preliminary results of PLSR models indicated that the inclusion of predictors other than total rainfall, percentage rainfall prior to 1 January, maximum temperature and soil type did not significantly improve prediction. The simulated weather data was contrasted with weather data based on weather station(s) records at or close to the simulated location using Agro-Climate Tool (Mauget & De Pauw, 2010).

Data Management

Observed and simulated data were examined for outliers, and to meet assumptions of univariate and multivariate normal distributions. Data transformations were performed if needed, then back-transformed for reporting. Annualized crop rotation yield was calculated, and two relational databases were constructed, one for observed data and one for simulated data. Covariates were added to each database where available and used in the statistical analyses (Lychuk et al., 2017). Yield gaps of annualized crop rotations due to RCP4.5 and RCP8.5, in comparison with RCP2.6, were estimated for each country, and the weighted averages for each country or location were used in the statistical analyses.

Results and Discussion

Simulation Results

Standardized biomass, grain, and soil C averaged over locations and countries under RCP4.5 (Figure 8.7a) and RCP8.5 (Figure 8.7b) and their basic statistics (i.e., mean deviation from overall average, standard deviation, and correlation coefficients) indicated a shift in magnitude in all three variables due to climate change. Overall distribution of biomass, grain, and soil C, in decreasing order, shifted below zero, whereas their variation decreased in magnitude. However, their relationships remained almost the same. Biomass and grain yield were strongly correlated under both RCPs, as expected; however, soil C expressed a smaller, but significant relationship with both variables.

Despite the long-term simulation modeling and the inclusion of diverse crop rotations, soil C declined in relative terms under RCP8.5. Obviously, the large impact of declining rainfall and higher maximum temperatures may have

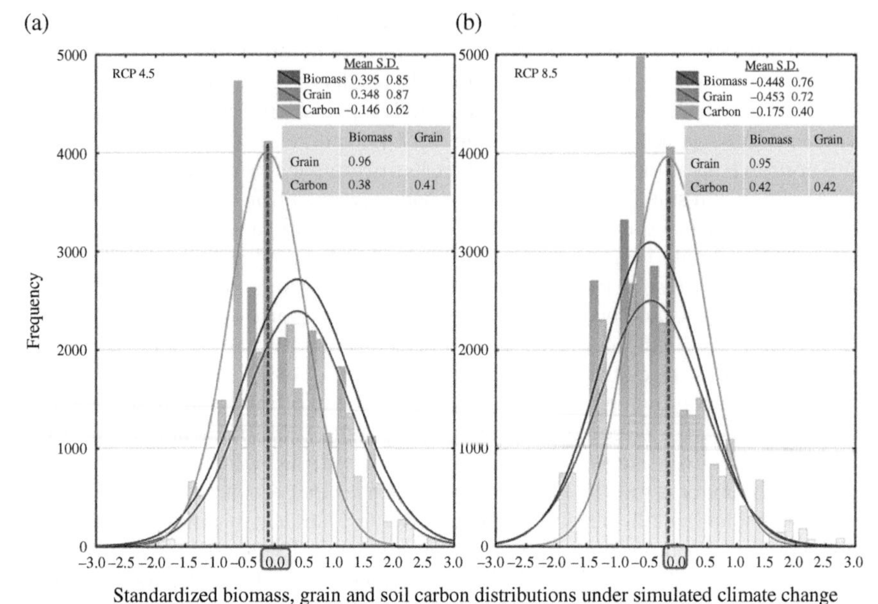

Figure 8.7 Probability distribution of simulated and standardized biomass, grain, and soil C averaged over 28 rain-fed locations across the Fertile Crescent under (a) RCP4.5 and (b) RCP8.5 representative concentration pathways.

contributed to this decline in soil C (Brilli et al., 2017). As a remedy for this potential scenario, the implementation of crop rotations with plants that allocate more C to top and subsoil through a deep root system is an adaptation measure for better C sequestration (Lagacherie et al., 2018).

Fixed factors and covariates had stronger effects on biomass and grain yield compared with their effects on soil C and soil N (Table 8.3) (Lychuk et al., 2017). Large geographic and local environmental and edaphic conditions, in line with significant differences in crop rotation yields with and among countries (Figure 8.8) masked any significant effects on all four variables, whereas differences between RCPs and soil series resulted in highly or marginally significant differences between the dependent variables. On the other hand, all covariates, in line with earlier results (Dixit & Telleria, 2015; Ghanem et al., 2015; Hirsch et al., 2018; Rolla et al., 2018), especially percentage rainfall prior to January, had significant effects (i.e., R^2 estimates) on all variables, except for the nonsignificant effects of maximum temperature on soil C and N (Table 8.3).

Random factors accounted for larger variances (Table 8.3) in biomass (82.1%) and grain yield (84.0%) than in soil C (67.8%) and soil N (47.9%). These differences are attributed to different amounts of variation accounted for by the individual random factors. Fifty percent of variances explained by differences among random factors were significant. However, the estimates ranged from small but significant 12.5% (biomass yield) due to locations (country), to large and highly significant for biomass (29.6%) and grain (23.2%) yields. These differences were

Table 8.3 Estimated effects of fixed factors (country, RCP and soils), covariates, and variance components of random factors (interacting and nested factors) on annualized biomass and grain yield of crop rotations and soil C in 28 rain-fed locations within seven countries in the Fertile Crescent

		Dependent variables			
Fixed factors	Random factors	Biomass yield	Grain yield	Soil C	Soil N
		p value			
Country		.12	.09	.12	.56
RCP		.001	.001	.05	.07
Soil series		.05	.05	.09	.07
Covariates					
		Coefficient of determination, R^2			
Total rainfall		.78**	.69**	.19	.15
		.01	.01	.05	.05
Rainfall before January		.82**	.75**	.25	.22
		.01	.01	.05	.05
Maximum temperature		.46*	.60**	.12	.09
		.05	.01	.10	.14
		Percentage variance (*z* significance)			
	Locations (country)	12.5*	18.9**	5.8	3.3
	RCP × Locations	18.6**	22.3**	16.2*	9.7
	RCP × Location (country)	29.6***	23.2***	12.8*	8.5
	Soils (country)	15.6**	12.9*	18.2*	12.7*
	RCP × Soils	3.5	4.0	5.6	5.5
	RCP × Soils (country)	1.9	2.7	9.2	8.2
Total variance		82.1	84.0	67.8	47.9

Note. RCP, representative concentration pathway.
* Significant at the .05 probability level.
** Significant at the .01 probability level.
*** Significant at the .001 probability level.

accounted for by the interaction between RCP and locations within countries. Soils, as part of random factors, showed limited effect on all dependent variables, except on soil N. It is speculated that differences in N (and other nutrients) endowment between calcareous soils and other soil series (Figure 8.6) may have contributed to this effect (Ryan & Sommer, 2012; Smith et al., 2016, 2017).

Cumulative probability of producing a given amount of annualized crop rotation yield under RCP4.5 and RCP8.5 compared with water-limited yield (Figure 8.9) indicated that on average (i.e., 50% cumulative probability), Y(RCP4.5) and Y(RCP8.5) would drop by 34 and 65%, respectively, compared with water-limited yield, Y_w; a similar trend was observed in yield (Y) variation, expressed as standard deviation. Relationships between the "normal" distribution, each of empirical cumulative distribution function (CDF), and the ±95.0% confidence ellipsis suggested that cumulative probability based on normal distribution was larger than the CDF up to 20% in the Y_w and Y(RCP4.5), but not in Y(RCP8.5). Variation around the mean, expressed as ±95% coefficient of variation [CV] were largest for RCP8.5, whereas for most of the probability distribution, empirical CDF was larger than the normal distribution, except for Y(RCP8.5) (Figure 8.9c). Simulation models predicted that regional effects of climate change will become more evident in the FC

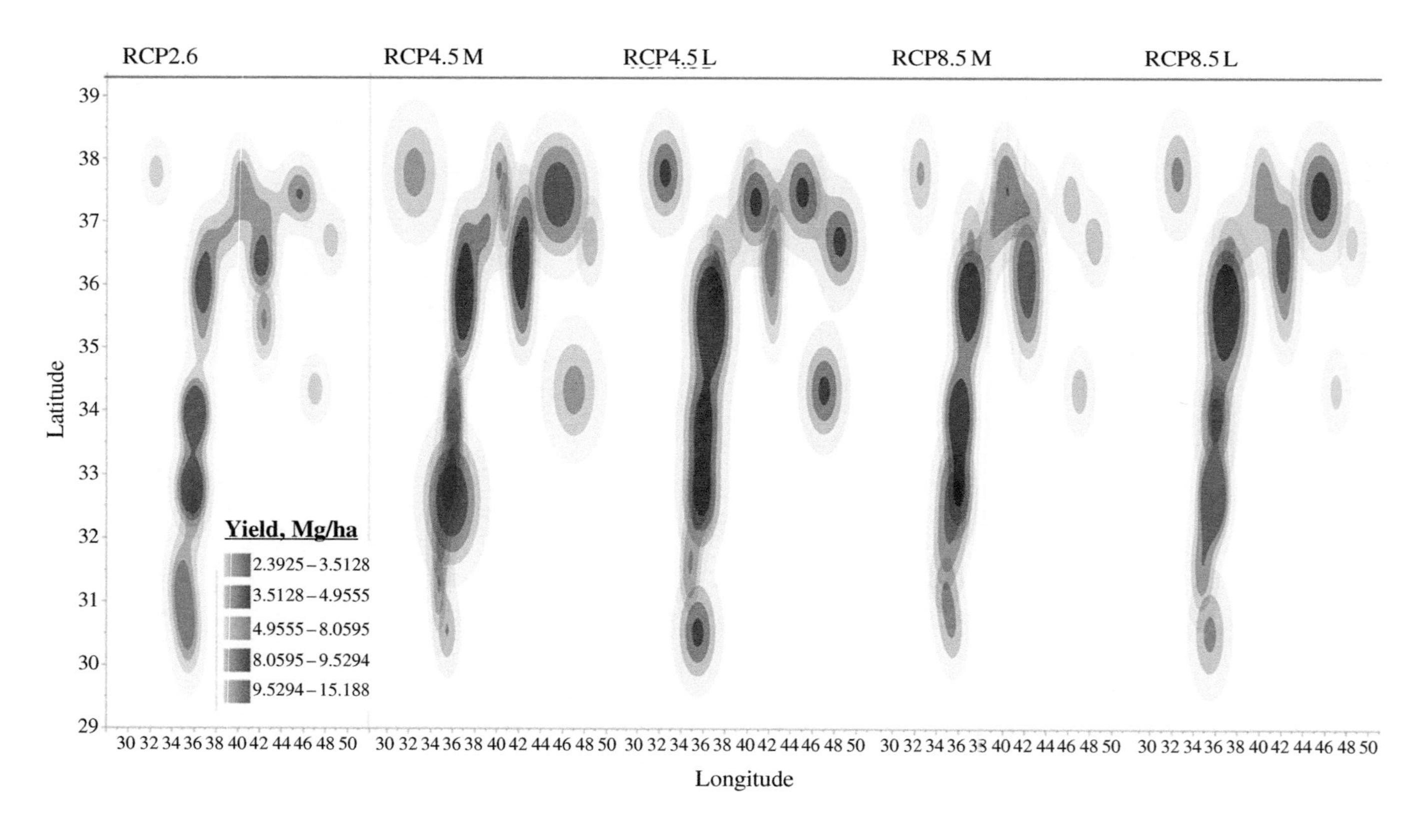

Figure 8.8 Heatmap of the spatial distribution of decreasing biomass yield averaged over annualized crop rotations in 28 rain-fed locations across the Fertile Crescent in response to medium- and long-term RCP4.5 and RCP8.5, in comparison with RCP2.6 representative concentration pathways.

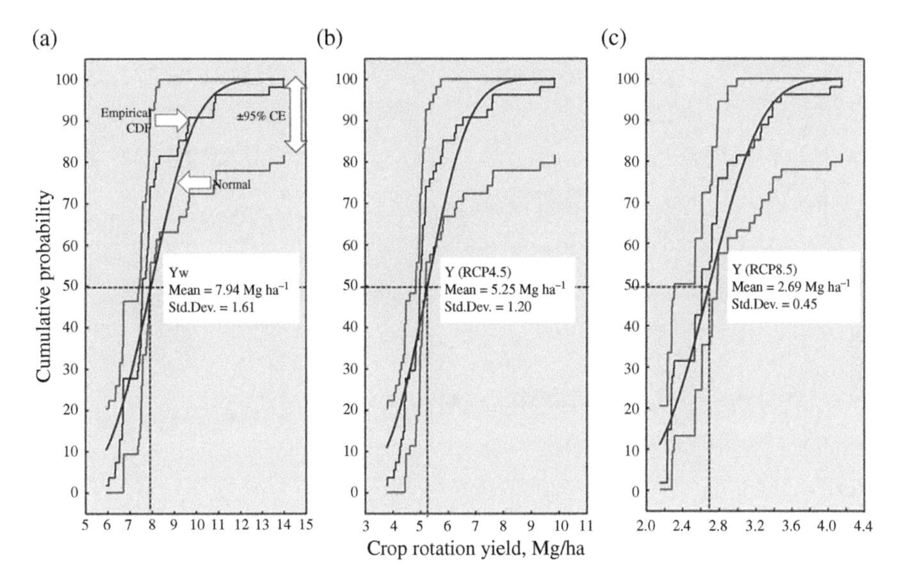

Figure 8.9 Cumulative probability distribution and its ±95% confidence ellipsis of (a) simulated water-limited biomass yield and yield in response to (b) RCP4.5 and (c) RCP8.5 representative concentration pathways averaged over annualized crop rotations in 28 rain-fed locations in rain-fed areas of the Fertile Crescent.

(and the larger Middle East region) in the absence of mitigation and adaptation measures; however, all countries will most likely experience 30 and 60% decline in crop yields with 1.5–2.0 and 3.0–4.0 °C of warming, respectively (Bassu et al., 2009; Sivakumar et al., 2013; Benton et al., 2018; Corbeels et al., 2018; Dixit et al., 2018).

The general progressive decline (from 15 to <2.5 Mg ha^{-1}, or almost 84.0%) in the estimated crop rotation yield along the longitude and latitude coordinates of the 28 locations in the FC under five RCPs (Figure 8.8) exceeded earlier estimates (Bassu et al., 2009; Benton et al., 2018; Corbeels et al., 2018; Dixit et al., 2018). The heatmaps indicated that location yields did not respond in a linear manner to changes in RCPs. In addition, variability in yield, as depicted by the range and shifts in heatmap, was large and dynamic in space and over time. Large differences, especially in middle and northern Levant (Casana, 2008), are observed between RCP2.6 and RCP4.5 M, followed by larger differences between the medium- and long-term RCP4.5 throughout the FC. The magnitude of yield decline between the medium- and long-term RCP8.5 was stronger at a few locations rather than at the larger scale of the FC. The fluctuations in frequency and rates of precipitation (Bassu et al., 2009), in addition to differences in soil native fertility (Ryan & Sommer, 2012), undoubtedly resulted in these yield discrepancies.

Current (RCP2.6) and future (RCP8.5) isohyets of simulated and annualized crop rotation yield, annual rainfall, and percentage annual rainfall before January (Figure 8.10) showed complex relationships between these variables across the FC. Compared with their densities and scales under RCP2.6, all three variables expressed shifts under RCP8.5. Largest shifts, especially in percentage rainfall before January, were observed, as expected in the Levant. It was postulated that as

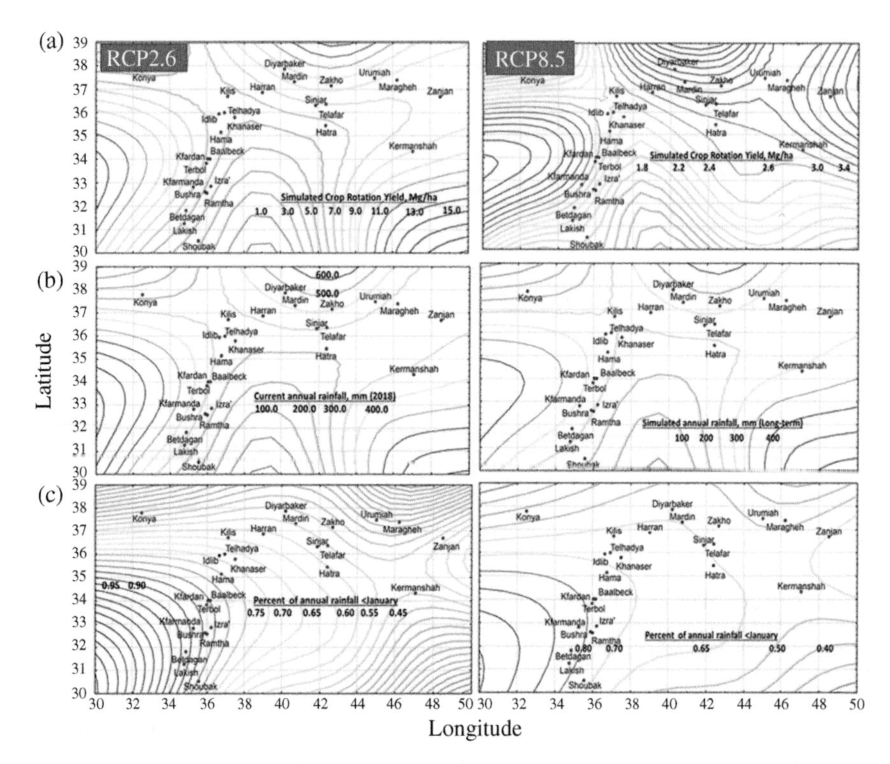

Figure 8.10 Isohyets of (a) simulated and annualized crop rotation yield, (b) current annual rainfall, and (c) percentage of rainfall prior to January in 28 rain-fed locations across the Fertile Crescent in response to RCP2.6 and RCP8.5 representative concentration pathways.

climate change progresses, average annual and seasonal temperatures in the Levant will rise and precipitation regimes will change by 2050 (Fridman & Kissinger, 2018). For example, isohyets of the southernmost location (Shoubak, Jordan) shifted from 3.0 Mg ha^{-1} (Figure 8.10a), 250 mm (Figure 8.10b), and 0.85 (Figure 8.10c) for yield, rainfall and percentage rainfall before January, respectively, under RCP2.6 to the corresponding values of 1.8, 210, and 0.7 under RCP8.5. On the other hand, respective values for a northern location (e.g., Harran, Turkey) were 8.0, 350, and 0.57 under RCP2.6, and 3.4, 280, and 0.65 under RCP8.5. For example (Lobell et al., 2008; Rolla et al., 2018), and despite temperature as the major driver of overall impacts in the Mediterranean climate, a reduction in the frequency of days with mean precipitation levels between 5 and 15 mm d^{-1} could be particularly detrimental to most crops in the Levant.

Statistical Modeling

Statistical modeling of crop, crop rotation, or cropping system yield using agrometeorological inputs into a statistical regression has been used in research programs (e.g., Lobell et al., 2008). A simple statistical model is built using a matrix with current or historic yield and several agrometeorological parameters (e.g., temperature and rainfall); then, a regression equation is derived between yields as a function of one or several agrometeorological variables. In the current study, the use of PLSR is statistically more appropriate than ordinary regression because it assumes that all variables are measured with error, and it takes into consideration the variation in crop rotation yield, rainfall, temperature, and soil, as well as the joint variation between these variables (SAS Institute, 2016; TIBCO Software, 2017). In addition, the validation variance provides another metric for model testing (R Core Team, 2018).

Monthly Rainfall

Distribution of rain events (e.g., monthly rainfall) plays a major role in determining dryland crop yields, whereas agricultural productivity in dryland regions is vulnerable to changes in precipitation patterns. This vulnerability increases with a decrease in total precipitation and is worse in regions where rainfall distribution is unimodal (e.g., the Levant). Sharp precipitation gradients are evident (Figure 8.10) across the longitude and latitude of the FC, where wheat yield, for example, is expected to decline by 38–52% (Balkovič et al., 2014; Haim et al., 2008). Variation in monthly rainfall (October–May) accounted for decreasing amounts of variation in crop rotation yield (Y) due to medium or long duration of RCP4.5 and RCP8.5 (Figure 8.11). The PLSR coefficients expressed large variation within and among RCPs. Monthly rainfall under RCP4.5 M and RCP4.5 L formed four factors and explained a total of 78.8 and 70.8% of simulated crop rotation variation, respectively. Contributions of individual months in explaining Y variance

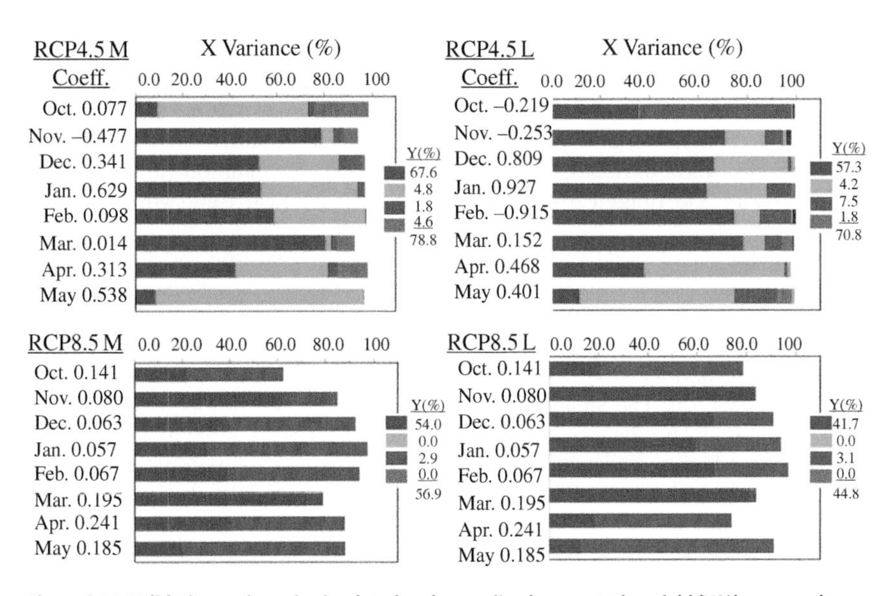

Figure 8.11 Validation variance in simulated and annualized crop rotation yield (Y%) averaged over crop rotations in 28 rain-fed locations across the Fertile Crescent, attributed to partial least square regression coefficients for the standardized and scaled rainfall data under combinations of medium- and long-term RCP4.5 and RCP8.5 representative concentration pathways.

were almost consistent under the medium and long duration of RCP4.5. In comparison, variation in monthly rainfall under medium- and long-term RCP8.5 formed only two factors and accounted for smaller amounts of variation in *Y* (56.9 and 44.3%, respectively). Unlike their values under RCP4.5, PLSR coefficients were much smaller under RCP8.5 and showed very little, if any, changes between medium- and long-term duration of RCP8.5. However, in all four cases, the first factor accounted for most variation in *Y*. Under relatively similar Mediterranean climate (Rolla et al., 2018), the beneficial effects of early rainfall (September–October) under RCP4.5 diminished with further reduction in predicted wheat yield under RCP8.5.

Variance Estimates

On average, the validation variance accounted for by four independent variables (Table 8.4) was relatively small and nonsignificant (27.6%) when compared with estimates based on individual RCPs or countries. In predicting the overall validation variance of crop rotation yield, coefficients of the PLSR model were positive for soils and total rainfall, but negative for percentage rainfall before January and maximum temperature. Larger and highly significant validation variances were obtained for all RCPs (47.2–79.6%); however, they expressed a decreasing trend as the level of climate stress increased from 79.7% under RCP2.6 to 47.2% under RCP8.5 L.

Independent variables differed as to the magnitude of their PLSR coefficients and variances when the validation variance of crop rotation yields was predicted under each RCP. Soils and total rainfall had positive coefficients; their variances ranged from 19.7–32.6% and 15.2–30.2%, respectively. Percentage rainfall prior to January had negative PLSR coefficients, whereas maximum temperature expressed larger variances than percentage rainfall before January (52.3–65.0, and 58.6–72.0, respectively). On the other hand, soils and total rainfall had positive PLSR coefficients, and percentage rainfall prior to January and maximum temperature effects were largely negative; nevertheless, validation variances were all significant and ranged from 56.7 to 70.9%, but no clear trend was observed across countries.

The coefficient of variation estimates differed in space (within and among countries) and time (under different RCPs), in addition to their variation due to different crop rotations (i.e., traditional vs. alternative; see Table 8.1). In general, CV estimates were larger in magnitude under alternative compared with traditional crop rotations and increased with increased intensity of climate change (i.e., from RCP2.6 to RCP8.5). The smallest CV estimates were found for Lebanon (23 and 25%) under RCP2.6, whereas the largest were found for Syria under RCP4.5 (43 and 47%) and for Turkey under RCP8.5 (46 and 63%); however, maximum differences between the smallest and largest estimates for each country had a relatively narrow range (from 14 to 19%), except for Turkey (27%).

Geographic and Agronomic Matrix Distances

The matrix correlation coefficient (Table 8.6) between geographical and RCP2.6 was significant ($r = .62$; $p = .05$), but not with RCP8.5 ($r = .45$; $p = .12$). In addition, the r estimates between RCP2.6 and RCP8.5 ($r = .37$; $p = .22$) was not significant.

Table 8.4 Validation variance in annualized crop rotation yield explained by soil series, annual rainfall, % rainfall before January, and maximum temperature during the growing season under RCP2.6, medium-, and long-term RCP4.5 and RCP8.5, and in each of seven countries in the Fertile Crescent

			Variance and standardized PLSR coefficients for explanatory variables				Crop rotation yield
Category	Factors	Model statistics	Soil series	Rainfall	% Rainfall before January	Maximum temperature (growing season)	Validation variance
All factors		Variance, %	16.8	29.8	30.0	79.4	27.6
		PLS coefficients	0.21	0.15	−0.21	−0.09	
RCP	RCP2.6	Variance, %	32.6	25.9	55.9	65.6	79.7***
		PLS coefficients	0.45	0.25	0.60	0.21	
	RCP4.5 M	Variance, %	29.2	28.2	65.0	58.6	62.8**
		PLS coefficients	0.27	0.18	−0.25	0.08	
	RCP4.5 L	Variance, %	22.7	30.2	52.6	61.2	60.9**
		PLS coefficients	0.27	0.95	−0.32	0.07	
	RCP8.5 M	Variance, %	19.6	15.2	63.2	58.3	57.8**
		PLS coefficients	0.31	0.32	−0.21	−0.50	
	RCP8.5 L	Variance, %	22.4	17.5	52.3	72.0	47.2*
		PLS coefficients	0.29	0.28	−0.17	−0.15	
Country	Iran	Variance, %	38.8	7.8	52.1	65.3	68.0**
		PLS coefficients	0.32	0.24	−0.48	−0.18	
	Iraq	Variance, %	45.0	65.0	15.6	62.0	56.7**
		PLS coefficients	0.42	0.51	−0.42	−0.42	
	Israel	Variance, %	48.7	35.0	52.6	72.3	66.9***
		PLS coefficients	0.48	0.32	−0.29	−0.21	
	Jordan	Variance, %	38.5	31.5	32.8	39.0	60.7**
		PLS coefficients	0.52	0.55	0.14	−0.29	
	Lebanon	Variance, %	59.0	51.2	32.0	63.2	70.9**
		PLS coefficients	0.48	0.35	−0.21	−0.19	
	Syria	Variance, %	39.6	82.0	5.8	62.8	57.5*
		PLS coefficients	0.27	0.47	0.11	−0.09	
	Turkey	Variance, %	43.8	61.0	18.0	52.2	65.9**
		PLS coefficients	0.56	0.49	−0.17	0.11	

Note. PLS, partial least squares; PLSR, partial least squares regression; RCP, representative concentration pathway.
* Significant at the .05 probability level.
** Significant at the .01 probability level.
*** Significant at the .001 probability level.

Geographical distances between countries in the FC expressed as D^2 between centroids of locations within each country ranged from 1.0 (maximum D^2 between Iran and Israel) to 0.25 between Israel and Lebanon. Five countries were 100% correctly classified, whereas Israel and Syria were 67 and 80% correctly classified, respectively, based on average crop rotation response to RCP8.5 (Table 8.5, diagonal). On the other hand, the multivariate distances between countries, on average, were four-fold smaller in response to RCP8.5 (mean $D^2 = 41.2$, above diagonal) compared with RCP2.6. (mean $D^2 = 160.0$, below diagonal). The majority of D^2 estimates under RCP2.6 (86.0%) were significant ($p < .05$); whereas the corresponding

Table 8.5 Coefficient of variation (CV%) of simulated and annualized crop rotation yield using traditional and improved management for the period 2010–2095 averaged over three to four rain-fed locations in each of seven countries of the Fertile Crescent under RCP2.6, RCP4.5, and RCP8.5

	Representative concentration pathways						
	RCP2.6		RCP4.5		RCP8.5		
	Crop rotations						Maximum
	Traditional	Alternative	Traditional	Alternative	Traditional	Alternative	difference
Country	–CV% of crop rotation yield–						%
Iran	28	30	29	34	36	42	14
Iraq	32	33	32	35	34	48	16
Israel	27	29	26	30	28	41	14
Jordan	34	38	40	42	44	53	19
Lebanon	23	25	31	33	35	38	15
Syria	40	45	43	47	42	59	19
Turkey	36	39	42	44	46	63	27
Avg.	31.4	34.2	34.7	37.9	37.9	49.2	17.7

Note. CV, coefficient of variation; RCP, representative concentration pathway.

value under RCP8.5 was 71%. Maximum D^2 values under RCP2.6 were between Iran and each of Israel (490) and Jordan (480); however, minimum values were between Syria and each of Jordan (14) and Israel (15). Maximum and minimum D^2 values under RCP8.5 were between Lebanon and each of Iran (105) and Jordan (9.0), respectively.

Yield Gaps

In addition to quantifying a potential yield, crop models have been used to understand the reasons for a yield gap, i.e., the difference between potential yield (Y_p) and yield under climate change (Lobell & Asseng, 2017; Neumann et al., 2010). Furthermore, assessments of climate change impacts on agricultural production, and the resulting yield gaps, may provide insight into how to sustainably intensify and diversify agriculture and optimize the use of natural resources and inputs, especially for subsistence farmers (Constantinidou et al., 2016; Fischer & Conner, 2018; Pretty & Bharucha, 2014).

The three-way relationships between potential yield (Y_p), and yield gaps (Y_g) under RCP4.5 and RCP8.5, separated the seven countries in the FC into three groups; Jordan and Iraq are expected to sustain the largest; Turkey, Lebanon and Iran the lowest; and Syria and Israel intermediate yield gaps (Figure 8.12). Yield gaps under RCP4.5 may range from 2.0 (e.g., in Turkey) to 5.0 Mg ha^{-1} (e.g., in Jordan), whereas those under RCP8.5 could be as high as 7.0–8.0 Mg ha^{-1} (in Syria, Jordan, and Iraq), but as low as 2.0–3.0 Mg ha^{-1} in Iran, Lebanon, and Turkey. The three-way relationship between Y_p, Y_g(RCP4.5), and Y_g(RCP8.5) was positive and significant ($r = .52; p = .05$). However, in assessing yield gaps, there is a spatial and temporal variability of agricultural land (Lobell & Asseng, 2017). Therefore, when yield gaps are evaluated (e.g., Neumann et al., 2010; van Bussel et al., 2015; http://www.yieldgap.org), inevitably there will be a discrepancy between actual yields, which are reported at the level of administrative units (e.g., water-limited yield, Y_w), and potential yield (Y_p). The latter is generally quantified at field level

Table 8.6 Correlation coefficients between matrices based on multivariate distances (Mahalanobis D^2) of geographic and crop rotation yields under RCP2.6 and RCP8.5 scenarios; and percentage correct classification (diagonal) of seven countries in the Fertile Crescent

Geographic D^2			Country	Iran	Iraq	Israel	Jordan	Lebanon	Syria	Turkey
						Geographical distances				
	Matrix r	p value								
			Iran	0.0[a]						
			Iraq	0.27	0.0[a]					
			Israel	1.00	0.89	0.0[a]				
			Jordan	0.95	0.64	0.32	0.0[a]			
	0.62	.05	Lebanon	0.86	0.78	0.25	0.36	0.0[a]		
			Syria	0.83	0.63	0.48	0.39	0.27	0.0[a]	
			Turkey	0.42	0.77	0.66	0.65	0.68	0.54	0.0[a]
						RCP2.6 distances				
RCP2.6			Iran	100[a]						
			Iraq	77*	100[a]					
			Israel	490***	230**	67[a]				
			Jordan	480***	220**	15	100[a]			
	0.45	.12	Lebanon	160**	53*	140**	140**	100[a]		
			Syria	440***	180**	15	14	130**	80[a]	
			Turkey	130**	32*	150**	150**	50*	108**	100[a]
						RCP8.5 distances				
RCP8.5			Iran	78[a]						
			Iraq	38*	54[a]					
			Israel	78*	69*	52[a]				
			Jordan	85*	23*	44*	67[a]			
	0.37	.22	Lebanon	105**	38*	36*	9	72[a]		
			Syria	50*	18	26*	13	18	83[a]	
			Turkey	15	21*	65*	18	69*	27*	85[a]
Geo. D^2										

Note. RCP, representative concentration pathway.
[a] These are correlation coefficients based on geographical "distances." The same countries have no distance.
* Significant at the .05 probability level.
** Significant at the .01 probability level.
*** Significant at the .001 probability level.

by using field trials, or point-based simulation models, similar to this and other studies referenced herein (e.g., Constantinidou et al., 2016; Dixit et al., 2018; Fischer & Conner, 2018; Ghanem et al., 2015; Soltani et al., 2016). Therefore, these yield gaps (Figure 8.12), averaged over annualized crop rotations and locations in each country, may not reflect yield gaps when assessed at the location level, or when using farm-based and experimental yield data. Nonetheless, closing these yield gaps demands immediate local adaptive research and will inevitably involve the adoption of management practices and inputs that has been developed and used elsewhere over the last century (Firbank et al., 2018; Fischer & Conner, 2018).

West Asia and North Africa, including the FC, and sub-Saharan Africa, were singled out (Fischer & Conner, 2018) as two major regions where yield gaps

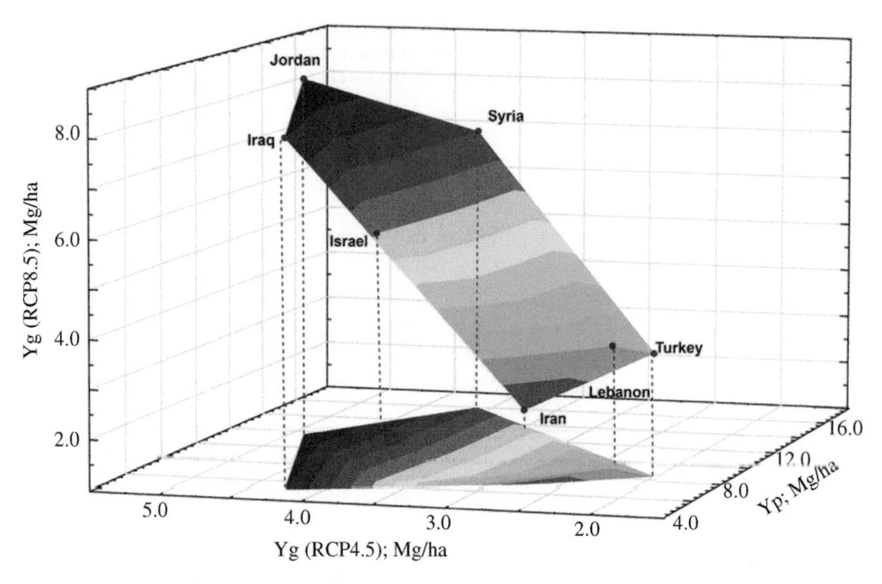

Figure 8.12 Yield gaps (Y_g; Mg ha^{-1}) due to RCP4.5 and RCP8.5 representative concentration pathways in comparison with yield potential (Y_p) averaged over annualized crop rotations in each of seven countries in the Fertile Crescent.

between farm yield and potential yield can be as high as 100 or even 200%. The main differences between traditional and improved crop rotations, and therefore, the yield gaps reported by several research papers and documented in the current study, were low inputs, short and simple crop rotations, and maximum and frequent tillage. Closing current pervasive yield gaps and moving the water-limited yield frontier (e.g., 2% per annum) may present an opportunity and a challenge to all involved, especially subsistence farmers in the FC (Fischer & Conner, 2018; Licker et al., 2010; van Oort et al., 2017). However, our understanding of the vulnerability and adaptive capacity of subsistence farming is limited; more quantitative and socioeconomic data is still needed.

Nonetheless, with 50% probability (Figure 8.9), subsistence farmers may be able to achieve as large as 67.0% (under RCP4.5) or as small as 34.0% (under RCP8.5) of their water-limited crop yield; whether these yield gaps can be closed agronomically, ecologically, and sustainably remains an open question (Firbank et al., 2018; Tittonell, 2014). However, to effectively strengthen food security of subsistence farmers in the FC, the larger the yield gap, the more urgent is the need to close it.

Conclusions

Landscapes of the FC have been the home for flourishing civilizations in antiquity; however, after being significantly degraded over time, they became the home for millions of smallholder subsistence dryland farmers who are vulnerable to climate change. These farmers face major challenges including decreasing and erratic rainfall, rising temperatures, and frequent droughts that limit the proper functioning and flow of agroecosystem services in subsistence farming. Our understanding of the vulnerability and adaptive capacity of these farmers remains limited and requires further data collection and quantitative research. Therefore, projections for the impact of climate change under RCP4.5 (at mid-century) and RCP8.5 (at end of century) scenarios on annualized yield and yield gaps of crop rotations were explored using simulation and statistical modeling. Under the three climate change scenarios, annualized crop rotation rain-fed yields were expected to decrease by 34 and 65% compared with current conditions. In spite of using crop rotations, soil C is also expected to decrease due to limited rainfall and increasing temperatures. The impacts of climate change on yields are expected to vary by region within the FC largely due to variation in precipitation.

It is expected to link yield gap analysis and crop rotation (and subsequently farming system) design in the subsistence farming regions of the FC by expanding the traditional yield gap analysis from individual crops to crop rotation level. Physiography (and geography) dictate, to some extent, the agroecological conditions and farming sustainability of subsistence farmers (Tittonell, 2014); this underscores the importance of geographical targeting and tailoring of interventions to increase their farm sustainability.

Despite the inherent uncertainty of predicting the impact of climate change on future crop yields, the study identified thresholds of technologies necessary for the proper functioning and flow of agroecosystem services in subsistence farming in seven countries of the FC. In addition, the impact of environmental variables and soils on the variability of soil C, crop yields, and yield gaps were quantified. The follow-up to this study are to (a) develop and test in silico strategies that target locations where yield of subsistence crop rotations are well below their current water-limited yield; (b) identify, at each location, strategies where crop rotation yields are closer to their current yield frontier; and (c) endorse the use of such strategies so that potential gains can be realized under subsistence farming in the FC.

Abbreviations

CMIP5, Coupled Model Intercomparison Project (fifth); CV, coefficient of variation; DSSAT, Decision Support System for Agrotechnology Transfer; FAO, Food and Agriculture Organization of the United Nations; FC, Fertile Crescent; GCMs, general circulation models; HAPPI-L; Half a degree Additional warming, Prognosis and Projected Impacts Land; IPCC, Intergovernmental Panel on Climate Change; RCP, representative concentration pathway; RMSE, Root Mean Square Error; Y_a, actual yield; Y_g, yield gap; Y_p, potential yield; Y_w, water-limited yield.

References

Abi Saab, M. T., Todorovic, M., & Albrizio, R. (2015). Comparing AquaCrop and CropSyst models in simulating barley growth and yield under different water and nitrogen regimes: Does calibration year influence the performance of crop growth models? *Agricultural Water Management, 147*, 21–33. http://doi.org/10.1016/j.agwat.2014.08.001

Adamo, N., Al-Ansari, N., Sissakian, V. K., Laue, J., & Knutsson, S. (2018). The future of the Tigris and Euphrates water resources in view of climate change. *Journal of Earth Sciences and Geotechnical Engineering, 8*, 59–74.

Adhikari, P., Araya, H., Aruna, G., Balamatti, A., Banerjee, S., Baskaran, P., Barah, B. C., Behera, D., Berhe, T., Boruah, P., Dhar, S., Edwards, S., Fulford, M., Gujja, B., Ibrahim, H., Kabir, H., Kassam, A., Khadka, R. B., Koma, Y. S., . . . Verma, A. (2018). System of crop intensification for more productive, resource-conserving, climate-resilient, and sustainable agriculture: Experience with diverse crops in varying agroecologies. *International Journal of Agricultural Sustainability, 16*, 1–28. https://doi.org/10.1080/14735903.2017.1402504

Alpert, P., Jin, F., & Kitoh, A. (2014). The projected death of the Fertile Crescent. In J. Norwine (Ed.), *A world after climate change and culture-shift* (pp. 193–203). Springer. https://doi.org/10.1007/978-94-007-7353-0_9

Asseng, S. (2013). Uncertainty in simulating wheat yields under climate change. *Nature Climate Change, 3*, 827–832. https://doi.org/10.1038/NCLIMATE1916

Balkovič, J., van der Velde, M., Skalský, R., Xiong, W., Folberth, C., Khabarov, N., Smirnov, A., Mueller, N. D., & Obersteiner, M. (2014). Global wheat production potentials and management flexibility under the representative concentration pathways. *Global and Planetary Change, 122*, 107–121. http://doi.org/10.1016/j.gloplacha.2014.08.010

Bassu, S., Asseng, S., Motzo, R., & Giunta, F. (2009). Optimizing sowing date of durum wheat in a variable Mediterranean environment. *Field Crops Research, 111*, 109–118. https://doi.org/10.1016/j.fcr.2008.11.002

Benli, B., Pala, M., Stöckle, C., & Oweis, T. (2007). Assessment of winter wheat production under early sowing with supplemental irrigation in a cold highland environment using CropSyst simulation model. *Agricultural Water Management, 93*, 45–53. https://doi.org/10.1016/j.agwat.2007.06.014

Benton, T. G., Bailey, R., Froggatt, A., King, R., Lee, B., & Wellesley, L. (2018). Designing sustainable land-use in a 1.5 °C world: The complexities of projecting multiple ecosystem services from land. *Current Opinion in Environmental Sustainability, 31*, 88–95. http://doi.org/10.1016/j.cosust.2018.01.011

Berry, E. M., Dernini, S., Burlingame, B., Meybeck, A., & Conforti, P. (2015). Food security and sustainability: Can one exist without the other? *Public Health Nutrition, 18*, 2293–2302. https://doi.org/10.1017/S136898001500021X

Braemer, F., Geyer, B., Castel, C., & Abdulkarim, M. (2010). Conquest of new lands and water systems in the western Fertile Crescent (central and southern Syria). *Water History, 2*, 91–114. https://doi.org/10.1007/s12685-010-0029-9

Brilli, L., Bechini, L., Bindi, M., Carozzi, M., Cavalli, D., Conant, R., Dorich, C. D., Doro, L., Ehrhardt, F., Farin, R., Ferrise, R., Fitton, N., Francaviglia, R., Grace, P., Iocola, I., Klumpp, K., Léonard, J., Martin, R., Massad, R. S., . . . Bellocchi, G. (2017). Review and analysis of strengths and weaknesses of agro-ecosystem models for simulating C and N fluxes. *Science of the Total Environment, 598*, 445–470. http://doi.org/10.1016/j.scitotenv.2017.03.208

Carranza-Gallego, G., Guzma, G. I., García-Ruíz, R., de Molina, M. G., & Aguilera, E. (2018). Contribution of old wheat varieties to climate change mitigation under contrasting managements and rainfed Mediterranean conditions. *Journal of Cleaner Production, 195*, 111–121. https://doi.org/10.1016/j.jclepro.2018.05.188

Casana, J. (2008). Mediterranean valleys revisited: Linking soil erosion, land use and climate variability in the northern Levant. *Geomorphology, 101*, 429–442. https://doi.org/10.1016/j.geomorph.2007.04.031

Constantinidou, K., Hadjinicolaou, P., Zittis, G., & Lelieveld, J. (2016). Effects of climate change on the yield of winter wheat in the eastern Mediterranean and Middle East. *Climate Research, 69*, 129–141. https://doi.org/10.3354/cr01395

Corbeels, M., Berre, D., Rusinamhodzi, L., & Lopez-Ridaura, S. (2018). Can we use crop modelling for identifying climate change adaptation options? *Agricultural and Forest Meteorology, 256–257*, 46–52. https://doi.org/10.1016/j.agrformet.2018.02.026

Cui, Z., Zhang, H., Chen, X., Zhang, C., Ma, W., Huang, C., Zhang, W., Mi, G., Miao, Y., Li, X., Gao, Q., Yang, J., Wang, Z., Ye, Y., Guo, S., Lu, J., Huang, J., Lv, S., Sun, Y., . . . Dou, Z. (2018). Pursuing sustainable productivity with millions of smallholder farmers. *Nature, 555*, 363–366. https://doi.org/10.1038/nature25785

Dixit, P. N., & Telleria, R. (2015). Advancing the climate data driven crop-modeling studies in the dry areas of northern Syria and Lebanon: An important first step for assessing impact of future climate. *Science of the Total Environment, 511*, 562–575. http://doi.org/10.1016/j.scitotenv.2015.01.001

Dixit, P. N., Telleria, R., Al Khatib, A. N., & Allouzi, S. F. (2018). Decadal analysis of impact of future climate on wheat production in dry Mediterranean environment: A case of Jordan. *Science of the Total Environment, 610–611*, 219–233. http://doi.org/10.1016/j.scitotenv.2017.07.270

FAO. (2018a). *FAO soils portal*. FAO. http://www.fao.org/soils-portal/soil-survey/soil-maps-and-databases/faounesco-soil-map-of-the-world/en

FAO. (2018b). *FAOSTAT*. FAO. http://www.fao.org/faostat/en/#data

Firbank, L. G., Attwood, S., Eory, V., Gadanakis, Y., Michael, J., Sonnino, R., & Takahashi, T. (2018). Grand challenges in sustainable intensification and ecosystem services. *Frontiers in Sustainable Food Systems, 2*, 7. https://doi.org/10.3389/fsufs.2018.00007

Fischer, R. A., & Conner, D. J. (2018). Issues for cropping and agricultural science in the next 20 years. *Field Crops Research, 222*, 121–142. https://doi.org/10.1016/j.fcr.2018.03.008

Folberth, C., Skalský, R., Moltchanova, E., Balkovič, J., Azevedo, L. B., Obersteiner, M., & van der Velde, M. (2016). Uncertainty in soil data can outweigh climate impact signals in global crop yield simulations. *Nature Communications, 7*, 1872. https://doi.org/10.1038/ncomms11872

Fridman, D., & Kissinger, M. (2018). An integrated biophysical and ecosystem approach as a base for ecosystem services analysis across regions. *Ecosystem Services, 31*, 242–254. https://doi.org/10.1016/j.ecoser.2018.01.005

Ghanem, M. E., Marrou, H., Soltani, A., Kumar, S., & Sinclair, T. R. (2015). Lentil variation in phenology and yield evaluated with a model. *Agronomy Journal, 107*, 1967–1977. https://doi.org/10.2134/agronj15.0061

Haim, D., Shechai, M., & Berliner, P. (2008). Assessing the impact of climate change on representative field crops in Israeli agriculture: A case study of wheat and cotton. *Climate Change, 86*, 425–440. https://doi.org/10.1007/s10584-007-9304-x

He, J., Jones, J. W., Graham, W. D., & Dukes, M. D. (2010). Influence of likelihood function choice for estimating crop model parameters using the generalized likelihood uncertainty estimation method. *Agricultural Systems, 103*, 256264.

Heng, L. K., Mejahed, K., & Rusan, M. (2007). Optimizing wheat productivity in two rain-fed environments of the West Asia–North Africa region using a simulation model. *European Journal of Agronomy, 26*, 121–129. https://doi.org/10.1016/j.eja.2006.09.001

Hirsch, A. L., Guillod, B. P., Seneviratne, S. I., Beyerle, U., Boysen, L. R., Brovkin, V., Davin, E. L., Doelman, J. C., Kim, H., Mitchell, D. M., Nitta, T., Shiogama, H., Sparrow, S., Stehfest, E., van Vuuren, D. P., & Wilson, S. (2018). Biogeophysical impacts of land-use change on climate extremes in low-emission scenarios: Results from HAPPI-land. *Earth's Future, 6*, 396–409. https://doi.org/10.1002/2017EF000744

Hole, F. (2009). Drivers of unsustainable land use in the semi-arid Khabur River Basin, Syria. *Geographical Research, 47*, 4–14. https://doi.org/10.1111/j.1745-5871.2008.00550.x

Hoogenboom, G., Jones, J. W., Wilkens, P. W., Porter, C. H., Boote, K. J., Hunt, L. A., Singh, U., Lizaso, J. I., White, J. W. Uryasev, O., Ogoshi, R., Koo, J., Shelia, V., & Tsuji, G. Y. (2015). *Decision Support System for Agrotechnology Transfer (DSSAT) Version 4.7.0*. DSSAT Foundation. http://dssat.net

Hoogenboom, G., Porter, C. H., Shelia, V., Boote, K. J., Singh, U., White, J. W., Hunt, L. A., Ogoshi, R., Lizaso, J. I., Koo, J., Asseng, S., Singels, A., Moreno, L P., & Jones, J. W. (2017). *Decision Support System for Agro-Technology Transfer (DSSAT) v. 4.7.0.0*. DSSAT Foundation. www.dssat.net

Intergovernmental Panel on Climate Change (IPCC). (2014). *Climate Change 2014. Synthesis Report*. Contribution of Working Groups I, II, and III to the Fifth Assessment Report of the Intergovernmental Panel on Climate Change. IPCC.

Jaradat, A. A. (2017). Agriculture in the Fertile Crescent: Continuity and change under climate change. *CAB Reviews, 12*, 1–33. https://doi.org/10.1079/PAVSNNR201712034

Jaradat, A. A., Riedell, W., & Goldstein, W. (2010). Biophysical constraints and ecological compatibilities of diverse agroecosystems. In E. Coudel, H. De-Vatour, C.-T. Soulard, & B. Hubert (Eds.), *Innovation and Sustainable Development in Agriculture and Food, Montpellier, 28–30 June*. Cirad–INRA–SupAgro. https://hal.archives-ouvertes.fr/hal-00510399

Johansen, C., Haque, M. E., Bell, R. W., Thierfelder, C., & Esdaile, R. J. (2012). Conservation agriculture for small holder rainfed farming: Opportunities and constraints of new mechanized seeding systems. *Field Crops Research, 132*, 18–32. https://doi.org/10.1016/j.fcr.2011.11.026

Jones, P. G., & Thornton, P. K. (2013). Generating downscaled weather data from a suite of climate models for agricultural modelling applications. *Agricultural Systems, 114*, 1–5. http://doi.org/10.1016/j.agsy.2012.08.002

Jones, J. W., He, J., Boote, K. J., Wilkens, P., Porter, C. H., & Hu, Z. (2011). Estimating DSSAT cropping system cultivar-specific parameters using Bayesian techniques. In L. R. Ahuja & L. Ma (Eds.), *Methods of introducing system models into agricultural research* (pp. 365–393). ASA, CSSA, and SSSA. https://doi.org/10.2134/advagricsystmodel2.c13

Kassie, B. T., Asseng, S., Porter, C. H., & Royce, F. S. (2016). Performance of DSSAT-N wheat across a wide range of current and future growing conditions. *European Journal of Agronomy, 81*, 27–36. http://doi.org/10.1016/j.eja.2016.08.012

Kelley, C. P., Mohtadi, S., Cane, M. A., Seager, R., & Kushner, Y. (2018). Climate change in the Fertile Crescent and implications of the recent Syrian drought. *Proceedings of the National Academy of Sciences of the USA, 112*, 3241–3246. https://doi.org/10.1073/pnas.1421533112

Kitoh, A., Yatagai, A., & Alpert, P. (2008). First super-high-resolution model projection that the ancient "Fertile Crescent" will disappear in this century. *Hydrological Research Letters, 2*, 1–4. https://doi.org/10.3178/HRL.2.1

Lagacherie, P. L., Álvaro-Fuentes, J., Annabi, M., Bernoux, M., Bouarfa, S., Douaoui, A., Grünberger, O., Hammani, A., Montanarella, L., Mrabet, R., Sabir, M., & Raclot, D. (2018). Managing Mediterranean soil resources under global change: Expected trends and mitigation strategies. *Regional Environmental Change, 18*, 663–675. https://doi.org/10.1007/s10113-017-1239-9

Lal, R. (2013). Climate change and soil quality in the WANA Region. In M. Sivakumar, R. Lal, R. Selvaraju, & I. Hamdan (Eds.), *Climate change and food security in West Asia and North Africa* (pp. 55–74). Springer. https://doi.org/10.1007/978-94-007-6751-5_3

Lancelotti, C. (2016). Resilience of small-scale societies' livelihoods: A framework for studying the transition from food gathering to food production. *Ecology and Society, 21*, 8. https://doi.org/10.5751/ES-08757-210408

Li, Z. T., Yang, J. Y., Drury, C. F., & Hoogenboom, G. (2015). Evaluation of the DSSAT-CSM for simulating yield and soil organic C and N of a long-term maize and wheat rotation experiment in the Loess Plateau of northwestern China. *Agricultural Systems, 135*, 90–104. http://doi.org/10.1016/j.agsy.2014.12.006

Li, Z., Yang, X., Cui, S., Yang, Q., Yang, X., Li, J., & Shen, Y. (2018). Developing sustainable cropping systems by integrating crop rotation with conservation tillage practices on the Loess Plateau, a long-term imperative. *Field Crops Research, 222*, 164–179. https://doi.org/10.1016/j.fcr.2018.03.027

Licker, R., Johnston, M., Foley, J. A., Bradford, C., Kucharik, C. K., Monfreda, C., & Ramankutty, N. (2010). Mind the gap: How do climate and agricultural management explain the 'yield gap' of crop lands around the world. *Global Ecology and Biogeography, 19*, 769–782. https://doi.org/10.1111/j.1466-8238.2010.00563.x

Lobell, D. B., & Asseng, S. (2017). Comparing estimates of climate change impacts from process-based and statistical crop models. *Environmental Research Letters, 12*(01500), 1.

Lobell, D. B., Burke, M. B., Tebaldi, C., Mastrandrea, M. D., Falcon, W. P., & Naylor, R. (2008). Prioritizing climate change adaptation needs for food security in 2030. *Science, 319*, 607–610. https://doi.org/10.1126/science.1152339

Lychuk, T. E., Moulin, A. P., Izaurralde, R. C., Lemke, R. L., Johnson, E. N., Olfert, O. O., & Brandt, S. A. (2017). Climate change, agricultural inputs, cropping diversity, and environmental covariates in multivariate analysis of future wheat, barley, and canola yields in Canadian prairies: A case study. *Canadian Journal of Soil Science, 97*, 300–318. https://doi.org/10.1139/cjss-2016-0075

Mahon, N., Curte, I., Simmons, E., & Islam, M. M. (2017). Sustainable intensification: "Oxymoron" or "third-way"? A systemic review. *Ecological Indicators, 74*, 73–97. https://doi.org/10.1016/j.ecolind.2016.11.001

Mahon, N., Curte, I., Di Bonito, M., Simmons, E. A., & Islam, M. M. (2018). Towards a broad and holistic framework of sustainable intensification indicators. *Land Use Policy, 77*, 576–597. https://doi.org/10.1016/j.landusepol.2018.06.009

Malek, Ž., & Verburg, P. (2017). Mediterranean land systems: Representing diversity and intensity of complex land systems in a dynamic region. *Landscape and Urban Planning, 165*, 102–116. http://doi.org/10.1016/j.landurbplan.2017.05.012

Malek, Ž., Verburg, P., Geijzendorffer, I. R., Bondeau, A., & Cramer, W. (2018). Global change effects on land management in the Mediterranean region. *Global Environmental Change, 50*, 238–254. https://doi.org/10.1016/j.gloenvcha.2018.04.007

Mauget, S., & De Pauw, E. (2010). The ICARDA agro-climate tool. *Meteorological Applications, 17*, 105–116. https://doi.org/10.1002/met.165

Michalscheck, M., Groot, J. C. J., Kotu, B., Hoeschle-Zeledon, I., Kuivanen, K., Descheemaeker, K., & Tittonell, P. (2018). Model results versus farmer realities. Operationalizing diversity within and among smallholder farm systems for a nuanced impact assessment of technology packages. *Agricultural Systems, 162*, 164–178. https://doi.org/10.1016/j.agsy.2018.01.028

Moeller, C., Pala, M., Manschadi, A. M., & Meinke, H. (2007). Assessing the sustainability of wheat-based cropping systems using APSIM: Model parameterization and evaluation. *Australian Journal of Agricultural Research, 2007*(58), 75–86. https://doi.org/10.1071/AR06186

Moeller, C., Sauerborn, J., de Voil, P., Manschadi, A. M., Pala, M., & Meinke, H. (2014). Assessing the sustainability of wheat-based cropping systems using simulation modelling: Sustainability. *Sustainability Science, 9*, 1–16.

Mrabet, R. (2011). No-tillage agriculture in West Asia and North Africa. In P. Tow, I. Cooper, I. Partridge, & C. Birch (Eds.), *Rainfed farming systems* (pp. 1015–1042). Springer. https://doi.org/10.1007/978-1-4020-9132-2_40

Mrabet, R., Moussadek, R., Fadlaoui, A., & van Ranst, E. (2012). Conservation agriculture in dry areas of Morocco. *Field Crops Research, 132*, 84–94. https://doi.org/10.1016/j.fcr.2011.11.017

Mutyasira, V., Hoag, D., Pendell, D., Manning, D. T., & Berh, M. (2018). Assessing the relative sustainability of smallholder farming systems in Ethiopian highlands. *Agricultural Systems, 167*, 83–91. https://doi.org/10.1016/j.agsy.2018.08.006

Nash, P. R., Gollany, H. T., Liebig, M. A., Halvorson, J. J., Archer, D. W., & Tanaka, D. L. (2018). Simulated soil organic carbon responses to crop rotation, tillage, and climate change in North Dakota. *Journal of Environmental Quality, 47*, 654–662. https://doi.org/10.2134/jeq2017.04.0161

Neumann, K., Verburg, P. H., Stehfest, E., & Muller, C. (2010). The yield gap of global grain production: A spatial analysis. *Agricultural Systems, 103*, 316–326. http://doi.org/10.1016/j.agsy.2010.02.004

O'Leary, C. J., Aggarwal, P. K., Calderini, D. F., Connor, D. J., Craufurd, P., Eigenbrode, S. D., Han, X., & Hatfield, J. L. (2018). Challenges and responses to ongoing and projected climate change for dryland cereal production systems throughout the world. *Agronomy, 8*, 34. https://doi.org/10.3390/agronomy8040034

Önol, B., Bozkurt, D., Turuncoglu, U. U., Lutfi Sen, O., & Nuzhet Dalfes, H. (2014). Evaluation of the twenty-first century RCM simulations driven by multiple GCMs over the eastern Mediterranean–Black Sea region. *Climate Dynamics, 42*, 1949–1965. https://doi.org/10.1007/s00382-013-1966-7

Pala, M., Oweis, T., Benli, B., De Pauw, E., El Mourid, M., Karrou, M., Jamal, M., & Zencirci, N. (2011). *Assessment of Wheat Yield Gap in the Mediterranean: Case Studies from Morocco, Syria and Turkey.* International Center for Agricultural Research in the Dry Areas (ICARDA).

Pretty, J., & Bharucha, Z. P. (2014). Sustainable intensification in agricultural systems. *Annals of Botany, 114*, 1571–1596. https://doi.org/10.1093/aob/mcu205

Qadir, M., Schubert, S., Oster, J. D., Sposito, G., Minhas, P. S., Cheraghi, S. A. M., Murtaza, G., Mirzabaev, A., & Saqib, M. (2018). High-magnesium waters and soils: Emerging environmental and food security constraints. *Science of the Total Environment, 642*, 1108–1117. https://doi.org/10.1016/j.scitotenv.2018.06.090

R Core Team. (2018). *R: A language and environment for statistical computing.* R Foundation for Statistical Computing. http://www.R-project.org

Reckling, M., Hecker, J.-M., Bergkvist, G., Watson, C. A., Zander, P., Schläfke, N., Stoddard, F. L., Eory, V., Topp, C. F. E., Maire, J., & Bachinger, J. (2016). A cropping system assessment framework: Evaluating effects of introducing legumes into crop rotations. *European Journal of Agronomy, 76*, 186–197. http://doi.org/10.1016/j.eja.2015.11.005

Riahi, K., Rao, S., Krey, V., Cho, C., Chirkov, V., Fischer, G., Kindermann, G., Nakicenovic, N., & Rafaj, P. (2011). RCP 8.5: A scenario of comparatively high greenhouse gas emission. *Climate Change, 109*, 33–57. https://doi.org/10.1007/s10584-011-0149-y

Riehl, S. (2016). The role of the local environment in the slow pace of emerging agriculture in the Fertile Crescent. *Journal of Ethnobiology, 36*, 512–534. https://doi.org/10.2993/0278-0771-36.3.512

Riehl, S., Pustovoytov, K. E., Weippert, H., Klett, S., & Hole, F. (2014). Drought stress variability in ancient Near Eastern agricultural systems evidenced by δ13C in barley grain. *Proceedings of the National Academy of Sciences of the USA, 111*, 12348–12353. https://doi.org/10.1073/pnas.1409516111

Rolla, A. L., Nuñez, M. N., Guevara, E. R., Meira, S. G., Rodriguez, G. R., & Záratea, M. I. O. (2018). Climate impacts on crop yields in Central Argentina. *Adaptation strategies. Agricultural Systems, 160*, 44–59. http://doi.org/10.1016/j.agsy.2017.08.007

Rosenzweig, C., Elliott, J., Deryng, D., Ruane, A. C., Müller, C., Arneth, A., Boote, K. J., Folberth, C., Glotter, M., Khabarov, N., Neumann, K., Piontek, F., Pugh, T. A. M., Schmid, E., Stehfest, E., Yang, H., & Jones, J. W. (2014). Assessing agricultural risks of climate change in the 21st century in a global gridded crop model intercomparison. *Proceedings of the National Academy of Sciences of the USA, 111*, 3268–3273. https:doi.org/10.1073/pnas.1222463110

Ruane, A. C., Phillips, M. M., & Rosenzweig, C. (2018). Climate shifts within major agricultural seasons for +1.5 and +2.0 °C worlds: HAPPI projections and AgMIP modeling scenarios. *Agricultural and Forest Meteorology, 259*, 329–344. https://doi.org/10.1016/j.agrformet.2018.05.013

Ryan, J., & Sommer, R. (2012). Soil fertility and crop nutrition research at an international center in the Mediterranean region: Achievements and future perspective. *Archives of Agronomy and Soil Science, 58*(Suppl. 1), S41–S54. https://doi.org/10.1080/03650340.2012.693601

Ryan, J., Singh, M., & Pala, M. (2008). Long-term cereal-based rotation trials in the Mediterranean Region: Implications for cropping sustainability. *Advances in Agronomy, 97*, 273–319.

Ryan, J., Ibrikci, H., Delgado, A., Torrent, J., Sommer, R., & Rashid, A. (2012). Significance of phosphorus for agriculture and the environment in the West Asia and North Africa Region. *Advances in Agronomy, 114*, 91–153. https://doi.org/10.1016/B978-0-12-394275-3.00004-3

Saïdi, S., Gintzburger, G., Gazull, L., Wallace, J., & Christiansen, S. (2018). A model for locating fodder shrub plantations sites in the Jordanian badiyah. *The Rangeland Journal, 40*, 527–538. https://doi.org/10.1071/RJ17129

Salah, Z., Nieto, R., Drumond, A., Gimeno, L., & Vicente-Serrano, S. M. (2018). A Lagrangian analysis of the moisture budget over the Fertile Crescent during two intense drought episodes. *Journal of Hydrology, 560*, 382–395. https://doi.org/10.1016/j.jhydrol.2018.03.021

SAS Institute. 2016. *JMP, version 13.2 Pro.* SAS Institute.

Saymohammadi, S., Zarafshani, K., Tavakoli, M., Mahdizadeh, H., & Amiri, F. (2018). Prediction of climate change induced temperature and precipitation: The case of Iran. *Sustainability, 9*, 146. https://doi.org/10.3390/su9010146

Shakoor, M. B., Ali, S., Tauqeer, H. M., Hannan, F., Rizvi, H., Ali, B., Iftikhar, U., & Farid, M. (2013). Vulnerability of climate change to agricultural systems in Third World Countries and adaptation practices. *International Journal of Agronomy and Plant Production, 4*, 2502–2513.

Silungwe, F. R., Graef, F., Bellingrath-Kimura, S. D., Tumbo, S. D., Kahimba, F. C., & Lana, M. A. (2018). Crop upgrading strategies and modelling for rainfed cereals in a semi-arid climate—A review. *Water, 10*, 356. https://doi.org/10.3390/w10040356

Sivakumar, M. V. K., Ruane, A. C., & Camacho, J. (2013). Climate change in the West Asia and North Africa region. In M. V. K. Sivakumar, R. Lal, R. Selvaraju, & I. Hamdan (Eds.), *Climate change and food security in West Asia and North Africa* (pp. 3–26). Springer. https://doi.org/10.1007/978-94-007-6751-5_1

Smith, A., Snapp, S., Dimes, J., Gwenambira, C., & Chikowo, R. (2016). Doubled-up legume rotations improve soil fertility and maintain productivity under variable conditions in maize-based cropping systems in Malawi. *Agricultural Systems, 145*, 139–149. http://doi.org/10.1016/j.agsy.2016.03.008

Smith, A., Snapp, S., Chikowo, R., Thorne, P., & Bekunda, M. (2017). Measuring sustainable intensification in smallholder agroecosystems. *Global Food Security, 12*, 127–138. http://doi.org/10.1016/j.gfs.2016.11.002

Soltani, A., Hajjarpour, A., & Vadez, V. (2016). Analysis of chickpea yield gap and water-limited potential yield in Iran. *Field Crops Research, 185*, 21–30. http://doi.org/10.1016/j.fcr.2015.10.015

Sommer, R., Piggin, C., Feindel, D., Ansar, M., van Delden, L., Shimonaka, K., Abdalla, J., Douba, O., Estefan, G., Haddad, A., Haj-Abdo, R., Hajdibo, A., Hayek, P., Khalil, Y., Khoder, A., & Ryan, J. (2014). Effects of zero tillage and residue retention on soil quality in the Mediterranean region of northern Syria. *Open Journal of Soil Science, 4*, 109–125. http://doi.org/10.4236/ojss.2014.43015

Sommer, R., Piggin, C., Haddad, A., Hajdibo, A., Hayek, P., & Khalil, Y. (2012). Simulating the effects of zero tillage and crop residue retention on water relations and yield of wheat under rainfed semiarid Mediterranean conditions. *Field Crops Research, 132*, 40–52. https://doi.org/10.1016/j.fcr.2012.02.024

Tadesse, W., Halila, H., Jamal, M., El-Hanafi, S., Assefa, S., Oweis, T., & Baum, M. (2017). Role of sustainable wheat production to ensure food security in the CWANA region. *Journal of Experimental Biology and Agricultural Sciences, 5*, 15–32. http://doi.org/10.18006/2017.5(Spl-1-SAFSAW).S15.S32

Tesfaye, W., & Seifu, L. (2016). Climate change perception and choice of adaptation strategies: Empirical evidence from smallholder farmers in East Ethiopia. *International Journal of Climate Change Strategies and Management, 8*, 253–270. https://doi.org/10.1108/IJCCSM-01-2014-0017

Thomson, A. M., Calvin, K. V., Smith, S. J., Kyle, G. P., Volke, A., Patel, P., Delgado-Arias, S., Bond-Lamberty, B., Wise, M. A., Clarke, L. E., & Edmonds, J. A. (2011). RCP4.5: A pathway for stabilization of radiative forcing by 2100. *Climate Change, 109*, 77–94. https://doi.org/10.1007/s10584-011-0151-4

Thornton, P. K., Whitbread, A., Baedeker, T., Cairns, J., Claessense, L., Baethgen, W., Bunn, C., Friedmann, M., Giller, K. E., Herrero, M., Howden, M., Kilcline, K., Nangia, V., Ramirez-Villegas, J., Kumar, S., West, P. C., & Keating, B. (2018). A framework for priority-setting in climate smart agriculture research. *Agricultural Systems, 167*, 161–175. https://doi.org/10.1016/j.agsy.2018.09.009

TIBCO Software. (2017). *Statistica (data analysis software system), version 13*. TIBCO Software. http://statistica.io

Tittonell, P. (2014). Ecological intensification of agriculture: Sustainable by nature. *Current Opinion in Environmental Sustainability, 8*, 53–61. http://doi.org/10.1016/j.cosust.2014.08.006

Tittonell, P., & Giller, K. E. (2013). When yield gaps are poverty traps: The paradigm of ecological intensification in African smallholder agriculture. *Field Crops Research, 143*, 76–90. http://doi.org/10.1016/j.fcr.2012.10.007

van Bussel, L. G. J., Grassini, P., Van Wart, J., Wolf, J., Claessens, L., Yang, H., Boogaard, H., de Groot, H., Saito, K., Cassman, K. G., & van Ittersum, M. K. (2015). From field to atlas: Upscaling of location-specific yield gap estimates. *Field Crops Research, 177*, 98–108. http://creativecommons.org/licenses/by-nc-nd/4.0

Van Duivenbooden, N., Pala, M., Studer, C., Bielders, C. L., & Beukes, D. J. (2000). Cropping systems and crop complementarity in dryland agriculture to increase soil water use efficiency: A review. *Netherlands Journal of Agricultural Science, 48*, 213–236.

van Oort, P. A. J., Saito, K., Dieng, I., Grassini, P., Cassman, K. G., & van Ittersum, M. K. (2017). Can yield gap analysis be used to inform R&D prioritization? *Global Food Security, 12*, 109–118. http://doi.org/10.1016/j.gfs.2016.09.005

van Vuuren, D. P., Edmonds, J., Kainuma, M., Riahi, K., Thomson, A., Hibbard, K., Hurtt, G. C., Kram, T., Krey, V., Lamarque, J.-F., Matsui, T., Meinshausen, M., Nakicenovic, N., Smith, S. J., & Rose, S. K. (2011). Representative concentration pathways: An overview. *Climatic Change, 109*, 5. https://doi.org/10.1007/s10584-011-0148-z

Waha, K., Krummenauer, L., Adams, S., Aich, V., Baarsch, F., Coumou, D., Fader, M., Hoff, H., Jobbins, G., Marcus, R., Mengel, M., Otto, I. M., Perrette, M., Rocha, M., Robinson, A., & Schleussner, C.-F. (2017). Climate change impacts in the Middle East and Northern Africa (MENA) region and their implications for vulnerable population groups. *Regional Environmental Change, 17*, 1623–1638. https://doi.org/10.1007/s10113-017-1144-2

Whitbread, A. M., Robertson, M. J., Carberry, P. S., & Dimes, J. P. (2010). How farming systems simulation can aid the development of more sustainable smallholder farming systems in southern Africa. *European Journal of Agronomy, 32*, 51–58. https://doi.org/10.1016/j.eja.2009.05.004

Yatagai, A. (2015). Trends in orographic rainfall over the Fertile Crescent. *Global Environmental Research, 15*, 147–156.

Zittis, G., Hadjinicolaou, P., & Lelieveld, J. (2014). Role of soil moisture in the amplification of climate warming in the eastern Mediterranean and the Middle East. *Climate Research, 59*, 27–37. https://doi.org/10.3354/cr01205

Index

Enhancing Agricultural Research and Precision Management for Subsistence Farming by Integrating System Models with Experiments, First Edition. Edited by Dennis J. Timlin and Saseendran S. Anapalli.

DOI: 10.1002/9780891183891.index

Index

Printed and bound by CPI Group (UK) Ltd, Croydon, CR0 4YY

16/04/2025

14658600-0001